Lecture Notes in Computer Science 11966

More information about this series at http://www.springer.com/series/7409

Makoto P. Kato · Yiqun Liu · Noriko Kando ·
Charles L. A. Clarke (Eds.)

NII Testbeds and Community for Information Access Research

14th International Conference, NTCIR 2019
Tokyo, Japan, June 10–13, 2019
Revised Selected Papers

 Springer

Editors
Makoto P. Kato
University of Tsukuba
Tsukuba, Japan

Yiqun Liu
Tsinghua University
Beijing, China

Noriko Kando
National Institute of Informatics (NII)
Tokyo, Japan

Charles L. A. Clarke
University of Waterloo
Waterloo, ON, Canada

ISSN 0302-9743 ISSN 1611-3349 (electronic)
Lecture Notes in Computer Science
ISBN 978-3-030-36804-3 ISBN 978-3-030-36805-0 (eBook)
https://doi.org/10.1007/978-3-030-36805-0

LNCS Sublibrary: SL3 – Information Systems and Applications, incl. Internet/Web, and HCI

This Springer imprint is published by the registered company Springer Nature Switzerland AG
The registered company address is: Gewerbestrasse 11, 6330 Cham, Switzerland

Preface

The 14th NTCIR Conference on Evaluation of Information Access Technologies was held during June 10–13, 2019, in Tokyo, Japan. NTCIR is a series of collective evaluation efforts designed to enhance research on diverse information access technologies, including, but not limited to, cross-language and multimedia information access, question-answering, text mining, and summarization, with an emphasis on East Asian languages such as Chinese, Korean, and Japanese, as well as English. Launched in the late 1997, NTCIR has attracted hundreds of research groups from well over 20 different countries and regions. Each NTCIR conference concludes the researchers' efforts over the course of about 18 months, in the form of official results and future work items.

NTCIR 2019 was host to 7 tasks that were coordinated in parallel by around 40 task organizers, who operated under the central coordination of the program co-chairs. A total of 87 research groups from 20 countries/regions participated in the NTCIR 2019 tasks, to compete and collaborate on a common ground, and thereby advance the state of the art.

Among the tasks, there were five core tasks which explored problems that were well-known in the fields of information access, as well as two pilot tasks, which aimed to address novel problems for which there are uncertainties as to how to evaluate them. The core tasks included: Lifelog Search (Lifelog-3); Open Live Test for Question Retrieval (OpenLiveQ-2), QA Lab for Political Information (QALab PoliINFO), Short Text Conversation (STC-3), and We Want Web (WWW-2). Meanwhile, CLEF/NTCIR/TREC Reproducibility (CENTRE) and Fine-Grained Numeral Understanding in Financial Tweet (FinNum) were organized as pilot tasks.

There was a total of 47 active participating teams which submitted both results and work report papers. At the conference, the Program Committee (PC) members, task organizers, and the participants had productive discussions about these works and invited a number of the papers to be submitted to the post-conference proceedings. After receiving the submissions, the PC worked with the task organizers to make acceptance decisions, and finally, they accepted 15 of them. Therefore, we believe that the papers in these post-conference proceedings can be regarded as a summarization of the great efforts and valuable findings in this round of NTCIR.

The conference and program chairs of NTCIR 2019 extend their sincere gratitude to all task organizers and participating teams in this round of NTCIR. We are also grateful to the PC for the great reviewing effort that guaranteed NTCIR 2019 could feature a quality set of core tasks and pilot tasks. We also thank Springer for supporting the publication of the post-conference proceedings.

October 2019

Makoto P. Kato
Yiqun Liu
Noriko Kando
Charles L. A. Clarke

Organization

General Chairs

Charles L. A. Clarke	University of Waterloo, Canada
Noriko Kando	National Institute of Informatics, Japan

Program Chairs

Makoto P. Kato	University of Tsukuba, Japan
Yiqun Liu	Tsinghua University, China

Conference Program Committee

Makoto P. Kato	University of Tsukuba, Japan
Yiqun Liu	Tsinghua University, China
Ben Carterette	University of Delaware, USA
Hsin-Hsi Chen	National Taiwan University, Taiwan
Tat-Seng Chua	National University of Singapore, Singapore
Nicola Ferro	University of Padova, Italy
Kalervo Järvelin	University of Tampere, Finland
Gareth J. F. Jones	Dublin City University, Ireland
Mandar Mitra	Indian Statistical Institute, India
Douglas W. Oard	University of Maryland, USA
Maarten de Rijke	University of Amsterdam, The Netherlands
Tetsuya Sakai	Waseda University, Japan
Mark Sanderson	RMIT University, Australia
Ian Soboroff	NIST, USA
Emine Yilmaz	University College London, UK

Post-conference Proceedings Program Committee

Makoto P. Kato	University of Tsukuba, Japan
Yiqun Liu	Tsinghua University, China
Cathal Gurrin	Dublin City University, Ireland
Hen-Hsen Huang	National Chengchi University, Taiwan
Yasutomo Kimura	Otaru University of Commerce, Japan
Tetsuya Sakai	Waseda University, Japan
Weinan Zhang	Harbin Institute of Technology, China
Takehiro Yamamoto	University of Hyogo, Japan
Min-Yuh Day	Tamkang University, Taiwan
Takeru Yokoi	Tokyo Metropolitan College of Industrial Technology, Japan

Jun Xu	Renmin University of China, China
Chung-Chi Chen	National Taiwan University, Taiwan
Alan Spark	AUTO1 Group, Germany
Min Zhang	Tsinghua University, China
Yasuhiro Ogawa	Nagoya University, Japan
Lin Li	Wuhan University of Technology, China
Maofu Liu	Wuhan University of Science and Technology, China
Masako Nomoto	Yahoo Japan Corporation, Japan
Kotaro Sakamoto	Yokohama National University, Japan
Hisashi Miyamori	Kyoto Sangyo University, Japan
Luke Gallagher	RMIT University, Australia
Minlie Huang	Tsinghua University, China
Pei Ke	Tsinghua University, China
Frank Hopfgartner	The University of Sheffield, UK
Andrew Yates	Max Planck Institute for Informatics, Germany
Kazutaka Shimada	Kyushu Institute of Technology, Japan
Feng Ji	DAMO Academy, Alibaba Group, China
Peng Zhang	Tianjin University, China
Tomohiro Manabe	Yahoo Japan Corporation, Japan
Madoka Ishioroshi	National Institute of Informatics, Japan
Minoru Sasaki	Ibaraki University, Japan
Chao-Chun Liang	Academia Sinica, Taiwan
Kang Xin	Tokushima University, Japan
Jiaxin Mao	Tsinghua University, China
Yuki Arase	Osaka University, Japan
Abderrahim Ait Azzi	Fortia Financial Solutions, France

Contents

Short Text Conversation

We Want Web

Fine-Grained Numeral Understanding in Financial Tweet

Lifelog Search

An Instant Approach with Visual Concepts and Query Formulation Based on Users' Information Needs for Initial Retrieval of Lifelog Moments

Tokinori Suzuki$^{(\boxtimes)}$ ⓘ and Daisuke Ikeda

Department of Informatics, Kyushu University, Fukuoka, Japan
suzuki.tokinori.070@s.kyushu-u.ac.jp,
daisuke@inf.kyushu-u.ac.jp

Abstract. Smart devices, such as smartphones and wearable cameras, have become widely used, and lifelogging with such gadgets has been recognized as a common activity. Since this trend produces a large amount of individual lifelog records, it is important to support users' efficient access of their personal lifelog archives. NTCIR Lifelog task series have studied the retrieval setting as a task called *Lifelog Semantic Access sub-task (LSAT)*. This task is that, given a topic of users' daily activity or events, e.g. "Find the moments when a user was eating any food at his/her desk at work", as a query, a system retrieves the relevant images of the moments from users' lifelogging records of their daily lives. Although, in the NTCIR conferences, interactive systems, which can utilize searchers' feedback in the retrieval process, have showed the higher performance than systems in automatic manner without users' feedback in the retrieval process, interactive systems rely on the quality of initial results, which can be seen as results of automatic systems. We envision that automatic retrieval that will be used in interactive systems later. In this paper, therefore, based on a principal that the system should be easy to implement for the later applicability, we propose a method scoring lifelog moments using only the meta information generated by publicly available pretrained detectors with word embeddings. Experimental results show the higher performance of the proposed method than the automatic retrieval systems presented in the NTCIR-14 Lifelog-3 task. We also show the retrieval can be further improved by about 0.3 of MAP with query formulation considering relevant/irrelvant writing about multimodal information in query topics.

Keywords: Lifelog · Multimodal data · Query formulation · Boolean queries · Word embedding

1 Introduction

Thanks to the recent progress of devices around us, such as wearable cameras, fitness trackers, smartphones and so on, lifelogging has been becoming a common activity because of convenience and easiness of such gadgets. Since this trend produces increasing amounts of individual lifelog records, it is important to support users'

© Springer Nature Switzerland AG 2019
M. P. Kato et al. (Eds.): NTCIR 2019, LNCS 11966, pp. 3–15, 2019.
https://doi.org/10.1007/978-3-030-36805-0_1

efficient access to their personal archives. NTCIR Lifelog task series [2–4] (NTCIR-14 Lifelog3 task this year) have studied the retrieval as their one of the tasks called *Lifelog Semantic Access sub-task (LSAT)*.

In the LSAT task, systems are required to retrieve a number of specific moments, which are defined as events or activities happened throughout a day, from lifeloggers' recordings. One of the events, for example, is the scenes of users' having breakfast. Specifically, given a query topic which represents an event or an activity in users' daily lives, e.g. "Find the moments when a user was eating any food at his/her desk at work", a system retrieves relevant moments from multimodal data of lifelog records; lifelogging images as the main resources with associated metadata such as location information, users' activity, the numbers of their steps in the moments and so on.

Some of the participants at NTCIR-13 Lifelog-2 task pointed out that the one of the major difficulties of the LSAT lies in that it requires understanding users' events/activities from lifelog images and the metadata [6, 15]. Instead of lifelog images themselves, the visual concepts generated by *Image Classification (IC)* are usually used for the retrieval. IC has been actively studied in this decade such as in ILSVRC on object recognition [12] and location classification [7, 17]. For example, IC may classify lifelog images into classifier's pre-defined classes, e.g. "office" for location or "banana", perhaps when a user was eating the food, for images of a user's eating breakfast. When the LSAT systems retrieve moments based on the visual concepts and the associated metadata. For the running example of the query, if IC labels are "office" and "banana", there are no overlap of the vocabularies between IC labels and the query text of a user's eating breakfast, which is called semantic gap [6, 15].

In NTCIR-14 Lifelog-3, interactive systems [1, 5, 10], which assume users in the retrieval process, i.e. systems can utilize the users' feedback for their results, showed higher performance than systems in automatic manner without assuming users in the retrieval [14]. Since a variety of multimodal data is available for the task, interactive approaches can be reasonable to automatically weigh appropriate lifelog features for scoring queried moments with helps of users' feedback using machine learning techniques. However, interactive systems rely on their initial retrieval results [1, 5, 6], which is equivalent to results of automatic systems, studying retrieval in automatic manner is beneficial for improving interactive systems because that leads increasing users' positive feedback on the relevant moments in results used in the later retrieval.

In this paper, we propose an automatic method with the aim of high applicability to the later retrievals with ease, thus the method only uses the information easily available, but it still can offer relatively good quality results served as initial results used in interactive search later. Based on the aim, the method only uses visual concepts of images generated by pre-trained classifiers, which are often publicly available, and scores moments with them in word embedding spaces. Experimental results show high performance of the method compared with the automatic retrieval systems presented in NTCIR-14 Lifelog-3 task. We also show the results can be further improved by query formulation considering relevant/irrelvant writing in query topics.

In the rest of paper, Sect. 2 briefly explains the related work. Section 3 explains NTCIR-14 Test collection. Section 4 introduces the proposed method. Section 5 shows the experiment and its outcome. Section 6 concludes the paper.

2 Related Work

Since interactive systems have succeeded in the LSAT task series, they are a promising approach for the retrieval because of availability of users' feedback [1, 5, 6]. Using the feedback, for example, the VCI2R system is able to effectively use the important features of multimodal data of lifelog records by modeling relevance judging process of users with *conditional random field* [6]. However, these results of the system influenced by the quality of the first retrieval which is equivalent to automatic retrievals.

Automatic systems, by contrast, may have a difficulty on how to utilize the multimodal data for the retrieval. This may be the reason many systems use visual information of lifelog images, e.g. objects, places or manually annotated activities, by visual concept detectors [6, 13–15]. But, there are two systems making use of the multimodality of the data for the task. First, PBG system applies a temporal filter to results for time-specific queries [15], though it considers only time-stamped information. Second, the baseline system at Lifelog-2 [18] translates a topic query to a Boolean sub-inquiry which designates each value of the types of lifelog data such as objects or activities. For example, given a query of "Find the moments a user was using laptop outside the work place", the translated inquiry designates the moments with "laptop" object and "working" activity. This system is quite close to our proposed method in this paper. However, the system uses term matching retrieval of the objective concepts, it may face the difficulty of the above-mentioned semantic gap. In addition, although they focused on manual translation of inquiries by assuming users in the retrieval process, we investigated the both manual and automatic ways of the translation.

3 Lifelog Test Collection

In this section, we explain the parts of NTCIR-14 Lifelog-3 test collection [4] that we used in this study. It consists of multimodal data of two users' (user 1 and user 2) lifelog records over 42 days in total. The multimodal data include information of blood glucose, calorie burn and heart rate. Though all of them may take part in the LSAT task, we selected the following multimodal data which may be useful for the task:

- Multimedia contents: Images captured by a wearable camera (OMG autographer[1]) put on the lifeloggers or manually taken by the users.
- Activity data: The activity of the lifeloggers captured by a smartphone application, Moves App[2]. The activity data are physical activities of the users such as walking, running and locations they visited.
- Biometric data: As mentioned above, we only used information of lifeloggers' steps recorded by the FitBit fitness trackers[3]. The steps information is expected to distinguish whether they were moving or not at that time.

---

[1] https://en.wikipedia.org/wiki/Autographer.

[2] http://www.moves-app.com/.

[3] https://www.fitbit.com/.

These multimodal data are timestamped, so that we can draw the data by IDs of minute as shown in Table 1. We used all the types of the data in the table.

Table 1. Multimodal data of lifelog records used in this study

Data	Explanation	Example
Minute ID	–	u1_20180503_0311
Local time	Date and time in local time zone	20180503_0411
Location	Location lifeloggers visited	Home, Dublin City University
Activity	Lifelogger's activity	Walking, transport
Steps	The # of steps in the minute	51
Linked image ID	Image IDs timestamped within the minute	u2_20180509_0933_i00

3.1   Images and the Visual Concepts

The test collection contains 64,132 and 17,615 lifelog images for user 1 and 2, respectively, during the period from 3 May 2018 to 31 May 2018. The images consist of two types: passive images and manual images. The passive images were captured by wearable cameras clipped to their clothing from the users' viewpoints from breakfast to evening. The manual images were conventional ones manually taken by the users.

The visual concepts of lifelog images generated by automatic detectors are provided in three types: Attribute, Category and Object. Figure 1 displays the examples. Note that the detectors employed publicly available pretrained models introduced in the following paragraph. Therefore, these visual concepts are readily available for the lifelog retrieval, and we used the information as a basis of our retrieval method.

	Attribute	Category	Cat. score	Object	Obj. score
	enclosed area	home office	0.595	chair	0.843
	indoor lighting	office	0.206	laptop	0.987
	studying	comput. room	0.076	keyboard	0.866
	...	...	...	NULL	–

Fig. 1. An example of visual concepts of a lifelog image

Attribute represents for environmental scenes and places of images. Category is information about semantic locations of images, e.g. home office and computer room. Attribute and Category were generated by *Places365-CNN* [17]. The last visual concept, Object is about objects detected in images. For example, a chair and laptop are the objects found in Fig. 1. To detect such objects, object detector trained on MSCOCO dataset [7] with *R-CNN* model [11] was used.

We note how to use the visual concepts. First, we used all of the ten Attributes for each image provided in the collection to cover both global aspects, e.g. enclosed area, and partial aspects, e.g. wood, of an image. Category is given as top five similar place labels for an image along with the probabilistic scores. Because the outputs of location change dramatically and we want to use the high reliability results, only the labels with the scores over a threshold, that we set at 0.2 for this time, were used. Lastly, for Object, because the recognition accuracy of R-CNN [11] is generally high, we used all the object labels as many as the recognizer found up to 25 labels each image.

3.2   Investigation on Topics of Lifelog Semantic Access Sub-Task (LSAT)

Because multimodal data in Table 1 are provided by the collection, we investigated applicability of the data for the retrieval by checking whether those are mentioned in the topics or not. For the investigation, we chose descriptions of the topics since they are assigned to all the topics, while narrative of topics is available for only two topics.

We observed that some of topics explicitly refer to the multimodal information of the finding moments in their descriptions; there are 20, 2 and 5 topics containing sentences about locations, time and activities of the queried moments, respectively over the 24 topics in total. We also found that 14 topics describe irrelevant moments in addition to the descriptions of relevant moments, which are in negative sentences.

As a result of the investigation, we got an idea that reflecting such multimodal information in queries which submitted to search systems, the retrieval results can be refined. For example, when a description refers to the moments happened in morning, the systems can omit the results in other time slots.

4   Lifelog Search with Query Formulation

In this section, we propose a retrieval method based on visual concepts of images using word embedding [8] for the LSAT task. First, we briefly explain word embedding. Then, we introduce how to adapt word embedding to the lifelog retrieval task. Finally, we describe query formulation to refine the results from retrieval with word embedding.

Places or objective concepts of images are easily obtained by pre-trained detectors, which are usually opened on the web. Our strategy of a lifelog search method is utilizing the visual concepts, which can be generated by such detectors. Thus, our method uses the text of the visual concepts as documents instead of images themselves.

We have already mentioned that if we adopt traditional term matching retrieval to the LSAT retrieval, the retrieval faces to the difficulty from a gap between vocabularies in the visual concepts and in queries, which is referred to as semantic gap [6] explained in Sect. 1. This term matching retrieval may result in poor search results.

To address the difficulty, we employed word embedding [8] for retrieval on the visual concepts motivated by recent successes of IR tasks, e.g. web search [9] and document retrieval in Software Engineering [16]. Word embeddings learn word vectors of real numbers that capture their contextual semantic meanings, such that similar words have similar vector representations. We expect the distributional nature of word vectors to measure the similarity between query terms and terms in the visual concepts of images, in which used vocabulary may be often different from each other.

4.1 Learning Word Embeddings

Many word embedding models have been proposed recently. Among them, *Continuous Bag-of-Words (CBOW)* and *Skip-gram* models are popular because of the software word2vec[4] that implements the models. As both models produce similar embeddings in quality and quantity, adopting the same learning policy in the web search work [9], we used CBOW model to learn a word embedding.

The model learns a word embedding by maximizing the log conditional probability of the word given the context words occurring within the fixed size of window around the word. Formally, let $c_k \in \mathbb{R}^d$ be a d-dimensional real number vector representing k th context word appearing in a $K - 1$ size window around a word w_i, which is represented by a vector $w_i \in \mathbb{R}^d$. The model predicts word w_i by adapting vector representation so that it has a large inner-product with the mean of the context word vectors. CBOW model minimizes the following objective:

$$\mathcal{L} = \sum_{i=1}^{|D|} - \log p(w_i|C_k)$$
$$= \sum_{i=1}^{|D|} - \log \frac{\exp\left(\overline{C}_K^\top w_i\right)}{\sum_{v=1}^{|V|} \exp\left(\overline{C}_K^\top w_v\right)},$$

where

$$\overline{C}_K = \frac{1}{K-1} \sum_{i-K \leq k \leq i+k, k \neq i} c_k$$

and D represents the training document collection. The probability is normalized by summing over all vocabularies, which is computationally expensive when training on large collection such as Wikipedia. Instead of the formulation, the model minimizes the following *negative sampling* objective:

$$- \log p(w_i|C_k) \approx - \log \sigma\left(\overline{C}_K^\top w_i\right) - \sum_{n=1}^{N} \log \sigma\left(\overline{C}_K^\top \hat{w}_n\right),$$

where N is the number of negative samples of words either from the uniform or empirical distribution over the vocabulary and σ is the sigmoid function.

---

[4] https://code.google.com/word2vec/.

4.2 Retrieving Images with Their Visual Concepts in Embedded Space

Since learnt embedding space contains distributional properties of word co-occurrence, word vectors would be a natural fit for modeling the aboutness of lifelog topics and visual concepts of images. With the embeddings, we determine the semantic related-ness of a query topic and visual concepts of image by computing the cosine similarity as a ranking function:

$$sim(Q, V) = \cos\left(\overline{Q}, \overline{V}\right) = \frac{\overline{Q}^\top \overline{V}}{\|\overline{Q}\|\|\overline{V}\|},$$

where

$$\overline{Q} = \frac{1}{|Q|} \sum_{q_i \in Q} \frac{q_i}{\|q_i\|} \text{ and } \overline{V} = \frac{1}{|V|} \sum_{v_j \in V} \frac{v_j}{\|v_j\|}.$$

$\overline{Q}$ and $\overline{V}$ are centroids of all the vectors for the terms in a query Q and visual concepts of an image V, respectively, each of which is a single embedding for the terms in a query or the visual concepts. Since word embeddings only apply to vocabulary they learnt, if terms are *out-of-vocabulary* words, the terms are ignored for the computing.

4.3 Query Formulation Based on Topic Descriptions

While using word embeddings by itself is effective for the LSAT task as we will demonstrate it in the experiments, the results can be refined by query formulation reflecting users' explicit information needs in topics because of the following two reasons.

First, the contribution of word embeddings would be on improving recall of retrieval results rather than precision of those due to much contextual information in the learnt vectors. That is, there may be quite lots of irrelevance in the results. Second, the search with word embeddings uses only visual information. However, there are other metadata of lifelog moments available, which is worth considering for the ranking because some of the topics denote the relation to the metadata such as locations, time or activities.

We propose two strategies to utilize the metadata by query formulation with sub-traction of query words on word embedding space or formalizing Boolean queries. Our aim for the query formulation is to reflect explicit information needs in the description parts of the topics into the searching queries.

Query Formulation with Vector Subtraction. The first way of query formulation is to use the characteristic of word embedding. The learnt word vectors by word embeddings contain some of linguistic regularities and patterns of words, and many of the patterns can be represented as linear translations, e.g. v ("Madrid") $-v$ ("Spain") $+v$ ("France") is the closer to v ("Paris") than any other vectors of words [8].

We adopt the linear translations to text of topics when generating query vectors. As we explained in Sect. 3, the fourteen topics clarify the irrelevant moments in their description part, such as "Being in an electronics store, or supermarket are not considered to be relevant" in a negative sentence in the description of topic 14006. Our assumption is that, if we subtract a word vector representing negative sentences in the description of a topic from the vector of whole the topic, the new vector fit more to the information needs in the topic. We formalize new queried vector $\overline{Q}^+$ as following equation.

$$\overline{Q}^+ = \overline{Q} - \overline{Q}^- ,$$

where $\overline{Q}$ and $\overline{Q}^-$ are the centroids of all the vectors for terms in whole sentences of a topic title or the description, and in negative sentences in the description, respectively.

Boolean Query Formulation. The second way of query formulation is Boolean queries. The topic descriptions denote not only irrelevant moments but also relevant ones mentioning locations where the querying events occurred or the time that happened, which can be useful for narrowing down the results of the visual concepts. We formulate Boolean queries in two manners, rule-based formulation and manual formulation.

For the rule-based formulation, first, we made three dictionaries of expressions of metadata in Data column of Table 1: time, locations and activities. The time dictionary contains expressions for time of a day, such as morning. The location contains places, e.g. home, hotel, store and so on, which are found in location data of the collection. The locations are grouped with their synonyms, e.g. hotel and inn are grouped with a synonym of hotel for normalization. The activity dictionary is only for expressions of walking and transport since only these two are provided in the collection. We linked words of transportation, such as trains and buses, to the transport as their synonym.

Second, we defined a simple translation rule to formulate Boolean queries. In the rule, our method finds phrases in the description matched to expressions in the dictionaries in affirmative sentences, and it formulates a logical AND operation which restricts the matched condition. For example, when the rule applied to the description of topic id 14012 "To be considered relevant, user 1 must be seen eating a scone with a cup of coffee in a hotel in the morning time", the left of query (1) "({06:00 AM-11:59 AM}[Time]" is generated when it find "morning" of the time dictionary in the description. When our method, by contrast, finds the phrases in negative sentences, it formulates a NOT operation for the condition. The method iterates the above step until it finds expressions of dictionaries in the description. Because the description of the running example of the topic also mentions that the location is in a hotel, a Boolean query for the topic is formulated as follows concatenating the operations:

$$(\{06:00\ AM\text{-}11:59\ AM\}[Time]\ AND\ \{Hotel,\ Inn\}[Location]) \tag{1}$$

The other way of manual formulation is conducted by users of the search system. They formulate Boolean queries manually using their knowledge. On topic 14024 finding the moments of user's just crossing the moments, the manual Boolean query is "NOT{0}[steps]" considering the user is walking at the moments, thus the number of steps does not equal to zero. Another example, on topic 14020 of finding the moments of the user in the café, thus the moments cannot be in home or in schools so that the Boolean query can be "NOT{Home, DCU}[Location]" where DCU means a university. Manual queries are formulated by users with these types of their inference.

5 Experiment

We evaluated the proposed method in the automatic retrieval setting on NTCIR-14 Lifelog-3 test collection described in Sect. 3. We used the LSAT relevance results generated by a pooled relevance judgement, which targets systems' output to assess relevance of topics, over the entire submissions from all the runs for the LSAT task as our ground-truth. If an approach retrieves one of the judged relevant results, we counted it as a correct. We did not make additional judgements for this evaluation based on the fact that automatic systems are generally weak systems than interactive systems, thus we assumed that the relevant results had been already discovered by the official results.

5.1 Evaluated Lifelog Retrieval Approaches

We evaluated three settings of the proposed method in Sect. 4 with variation of query formulation. *Concept* setting retrieves using visual concepts of images only. Concept setting is used with combination of each of the remaining two settings: *Concept + word2vec (word2vec)* and *Concept + Boolean (Boolean)*. The word2vec settings use queries generated by subtraction of word vectors from negative sentences in topic descriptions. The Boolean uses Boolean queries representing the descriptions.

The queries used in the word2vec settings or the Boolean settings are generated by either defined rules automatically or manual formulation. We examined settings of manual query formulations because we were interested in how high performances can be obtained when queries are manually fine-tuned with users' knowledge, that are seen as the upper-bounds. The details of an example of manually generated queries used in the experiments are shown in Table 2. All of the used queries are not present here due to the space limitation. They are uploaded to and opened on a website[5] instead.

For word2vec settings, although the rule-based subtraction uses all the terms in underlined sentence including "café" and "alone" on the topic 14020 in the table, manual formulation selects terms in the negative sentences for the subtraction. For example, the manual query formulation uses only "alone" because the topic finds the scenes in cafés, and subtraction with the vector of "café" may cause negative effects.

For manual Boolean query formulation, we generate queries with knowledge about our understanding of topics as users in a search process. For example, if a queried scene is about walking moments, the number of a lifelogger's steps at the moment would be zero. We add the restriction in the query. As another example of the knowledge, in case of finding scenes in a café, we exclude the interior scenes, e.g. home or school in Table 2 probably looks similar to café considering the accuracy of the used detectors.

---

[5] https://zenodo.org/record/3445638.

The word embedding was trained with skip-gram model at five window size on English version of Wikipedia dump data on October 2017. We use the learnt 300 dimensional vectors for the words.

Table 2. Examples of queries formulated from the relevant/irrelevant parts of the description. An underlined sentence writes about irrelevant scenes (a negative sentence).

Topic ID	A part of topic description	Boolean (rule)	Boolean (manual)	Word2vec (manual)
14020	To be considered relevant, u1 must be in a café and having coffee with another individual. Moments which show u1 alone in a café are not considered relevant.	({café} [location])	(({café} [location]) OR (NOT {Home, DCU}[Location]))	alone

5.2 Results

Table 3 shows the mean average precision (MAP), the precision at 5th, 10th and 30th (P@5, P@10 and P@30, respectively) of all the topics. The Boolean query settings achieved the best results at MAP of 0.308 with manual formulation, and the second best at that of 0.256 with the rule-based formulation.

All the results of examined settings are higher than those of automatic systems, QUIK and nlg 301, presented in the Lifelog-3 task. The Concept (Title) contributed the major part of this success showing MAP of 0.201 by itself. We checked two query setting of using the title or the description part of topics. The results of using title is higher than those of using description. On topic 14017, for example, the title is "Find the moments when user 1 was looking inside a refrigerator at home". The description of the topic additionally explains the irrelevant moments such as "Moments when eating food or cooking in the kitchen are not considered relevant" whose terms, such as "food" or "kitchen", can affect negatively in the similarity computing. Concept (Title), therefore, is used in the word2vec and the Boolean query settings. Both of the Boolean and the word2vec setting succeeded to improve the results of Concept setting.

Since the difficulty of LSAT task varies according to topics, we analyzed the detailed results of each topic in Table 4. There are 19 colored topics in the table where at least one of the query formulation can be applied to. On 16 out of the 19 topics, one of the settings could obtain higher AP than Concept only. In the results, when queries of manual formulation are not different from those of the rules, e.g. topic 14002, we adopted the results of rule-based formulation. Among them, Manual Boolean is the most successful setting achieving the best AP on eight topics. Therefore, the query formulations are generally effective.

For analysis of failures, the AP of the rule-based Boolean are lower than Concept only on some of the topics. A notable case is topic 14007 that is designed to find moments in a hotel. The Boolean query that restricts the moments with location

An Instant Approach with Visual Concepts and Query Formulation 13

information of hotels affected negatively. We observed that some of the images shot by
wearable cameras are blurred and unclear, or difficult to distinguish the locations, it
would be hard to precisely assign locations to the images. For these topics, using rules
that do not exclude the locations is one of the solutions.

Table 3. Retrieval performance of each settings. Figures indicate the average of all the topics.

Method	MAP	P@5	P@10	P@30	# of relevant results
QUIK Run 1	0.056	0.158	0.158	0.118	232
nlg 301 Run 1	0.063	—	0.238	—	293
Concept (Title)	0.202	0.167	0.192	0.172	565
Concept (Description)	0.138	0.067	0.092	0.133	400
Concept + *word2vec* (Sentence)	0.205	0.117	0.142	0.165	483
Concept + *word2vec* (Manual)	0.213	0.142	0.133	0.163	562
Concept + Boolean (Rule-based)	0.256	0.158	0.179	0.158	419
Concept + Boolean (Manual)	**0.308**	**0.242**	**0.242**	**0.244**	572

Table 4. Results on each topic. The best result of each topic shows in bold figures. Colored
background indicates the results obtained by the corresponding query formulation.

Topic id	Concept (Title) AP	P@10	+ word2vec (Rule) AP	P@10	+ word2vec (Manual) AP	P@10	+ Boolean (Rule) AP	P@10	+ Boolean (Manual) AP	P@10
14001	**0.239**	**0.400**	0.147	0.100	0.049	0.000	0.239	0.400	0.239	0.400
14002	0.332	0.200	0.197	0.200	0.198	0.300	**0.354**	**0.200**	0.354	0.200
14003	0.020	0.000	0.000	0.000	0.302	0.100	**1.000**	**0.100**	1.000	0.100
14004	**0.095**	**0.000**	0.095	0.000	0.095	0.000	0.095	0.000	0.095	0.000
14005	0.218	0.100	1.000	0.100	**0.505**	0.100	0.267	0.400	0.348	**0.500**
14006	0.078	0.100	0.016	0.000	0.016	0.000	**0.163**	**0.100**	0.163	0.100
14007	0.814	0.500	0.814	0.500	0.814	0.500	0.075	0.100	**0.908**	**0.800**
14008	0.035	0.000	0.037	0.000	0.278	0.100	0.325	**0.500**	**0.370**	0.400
14009	0.012	0.000	0.010	0.000	0.000	0.000	0.091	0.000	**0.109**	**0.100**
14010	**0.396**	**0.500**	0.312	0.400	0.075	0.100	0.102	0.000	0.290	0.300
14011	**0.507**	**0.600**	0.507	0.600	0.507	0.600	0.507	0.600	0.507	0.600
14012	0.300	0.400	0.363	0.500	**0.735**	0.600	0.687	**0.700**	0.687	0.700
14013	**0.164**	**0.200**	0.164	0.200	0.164	0.200	0.129	0.000	0.129	0.000
14014	0.050	0.100	0.053	0.100	**0.197**	0.100	0.142	**0.200**	0.142	0.200
14015	**0.045**	0.000	0.045	0.000	0.045	0.000	0.045	0.000	0.045	0.000
14016	**0.026**	0.000	0.026	0.000	0.026	0.000	0.026	0.000	0.026	0.000
14017	0.107	**0.100**	0.043	0.000	0.043	0.000	0.107	**0.100**	**0.138**	0.000
14018	0.759	**0.700**	0.772	0.400	0.772	0.400	**1.000**	0.100	0.790	0.500
14019	0.059	0.000	0.008	0.000	0.000	0.000	0.059	0.000	**0.016**	0.000
14020	0.141	0.200	0.125	**0.200**	0.089	0.000	0.041	0.000	**0.207**	0.200
14021	**0.034**	0.000	0.034	0.000	0.034	0.000	0.034	0.000	0.034	0.000
14022	**0.282**	**0.400**	0.020	0.000	0.048	0.000	0.275	0.400	0.275	**0.400**
14023	0.000	0.000	0.000	0.000	0.000	0.000	0.250	0.300	**0.272**	0.300
14024	0.129	0.100	0.129	0.100	0.129	**0.100**	0.129	0.100	**0.248**	0.000

6   Conclusion

In this paper, we proposed a retrieval method for Lifelog Semantic Access Task using visual concepts of images generated by publicly available pre-trained detectors so that it can be easily implemented and offer initial results for the later retrievals. The method ranks moments with the combinations of the visual concepts in word embedding space and query formulation. Experimental results showed the higher retrieval performance of the proposed method than automatic retrieval systems presented in NTCIR-14 Lifelog-3 task. In addition, query formulation is effective to improve the method with both of the rule-based formulation and manual formulation.

The manual formulation of Boolean queries was the best setting as we set this as the upper-bound setting. To make the rule-based settings close to the manual results, we need to build and arrange knowledge used in the formulation. For example, when given a query of an event happed outdoors, then formulating a query that exclude the location such as a supermarket recalling indoor situations. This is a future work of the study.

References

1. Fu, M.-H., Chang, C.-C., Huang, H.-H., Chen, H.-H.: Incorporating external textual knowledge for life event recognition and retrieval. In: Proceedings of the 14th NTCIR Conference on Evaluation of Information Access Technologies, pp. 61–71 (2019)
2. Gurrin, C., Joho, H., Hopfgartner, F., Zhou, L., Albatal, R.: Overview of NTCIR-12 lifelog task. In: Proceedings of the 12th NTCIR Conference on Evaluation of Information Access Technologies, pp. 354–360 (2016)
3. Gurrin, C., et al.: Overview of NTCIR-13 lifelog-2 task. In: Proceedings of the 13th NTCIR Conference on Evaluation of Information Access Technologies, pp. 6–11 (2017)
4. Gurrin, C., et al.: Overview of NTCIR-14 lifelog-3 task. In: Proceedings of the 14th NTCIR Conference on Evaluation of Information Access Technologies, pp. 14–26 (2019)
5. Le, N.-K., et al.: HCMUS at the NTCIR-14 lifelog-3 task. In: Proceedings of the 14th NTCIR Conference on Evaluation of Information Access Technologies, pp. 48–60 (2019)
6. Lin, J., Garcia del Molino, A., Xu, Q., Fang, F., Subbaraju, V., Lim, J.: VCI2R at NTCIR-13 lifelog-2 lifelog semantic access task. In: Proceedings of the 13th NTCIR Conference on Evaluation of Information Access Technologies, pp. 28–32 (2017)
7. Lin, T.-Y., et al.: Microsoft COCO: common objects in context. In: Fleet, D., Pajdla, T., Schiele, B., Tuytelaars, T. (eds.) ECCV 2014. LNCS, vol. 8693, pp. 740–755. Springer, Cham (2014). https://doi.org/10.1007/978-3-319-10602-1_48
8. Mikolov, T., Sutskever I., Chen, K., Corrado, G., Dean, J.: Distributed representations of words and phrases and their compositionality. In: Proceedings of the 26th International Conference on Neural Information Processing Systems, pp. 3111–3119 (2013)
9. Mitra, B., Nalisnick, E., Craswell, N., Caruana, R.: A dual embedding space model for document ranking. arXiv preprint arXiv:1602.01137 (2016)
10. Ninh, V.-T., et al.: A baseline interactive retrieval engine for the NTCIR-14 lifelog-3 semantic access task. In: Proceedings of the 14th NTCIR Conference on Evaluation of Information Access Technologies, pp. 72–80 (2019)
11. Ren, S., He, K., Girshick, R., Sun, J.: Faster R-CNN towards real-time object detection with region proposal networks. IEEE Trans. Pattern Anal. Mach. Intell. **39**(6), 1137–1149 (2017)

12. Russakovsky, O., et al.: ImageNet large scale visual recognition challenge. Int. J. Comput. Vision **115**(3), 211–252 (2015)
13. Safadi, B., Mulhem, P., Quénot G., Chevallet, J.-P.: LIG-MRIM at NTCIR-12 lifelog semantic access task. In: Proceedings of the 12th NTCIR Conference on Evaluation of Information Access Technologies, pp. 361–365 (2016)
14. Suzuki, T., Ikeda, D.: Smart lifelog retrieval system with habit-based concepts and moment visualization. In: Proceedings of the 14th NTCIR Conference on Evaluation of Information Access Technologies, pp. 40–47 (2019)
15. Yamamoto, S., Nishimura, T., Akagi, Y., Takimoto, Y., Inoue, T., Toda, H.: PBG at NTCIR-13 lifelog-2 lat, lsat, and lest tasks. In: Proceedings of the 13th NTCIR Conference on Evaluation of Information Access Technologies, pp. 12–19 (2017)
16. Ye, X., Shen, H., Ma, X., Bunescu, R., Liu, C.: From word embeddings to document similarities for improved information retrieval in software engineering. In: Proceedings of the 38th IEEE International Conference on Software Engineering, pp. 404–415 (2016)
17. Zhou, B., Lapedriza, A., Khosla, A., Oliva, A., Torralba, A.: Places: a 10 million image database for scene recognition. IEEE Trans. Pattern Anal. Mach. Intell. **40**(6), 1452–1464 (2017)
18. Zhou, L., Dang-Nguyen, D.-T., Gurrin, C.: A baseline search engine for personal life archives. In: Proceedings of the 2nd Workshop on Lifelogging Tools and Applications, pp. 21–24 (2017)

Advances in Lifelog Data Organisation and Retrieval at the NTCIR-14 Lifelog-3 Task

Cathal Gurrin[1]([✉]), Hideo Joho[2], Frank Hopfgartner[3], Liting Zhou[1],
Van-Tu Ninh[1], Tu-Khiem Le[1], Rami Albatal[1], Duc-Tien Dang-Nguyen[4],
and Graham Healy[1]

[1] Dublin City University, Dublin, Ireland
cathal.gurrin@dcu.ie
[2] University of Tsukuba, Tsukuba, Japan
[3] University of Sheffield, Sheffield, UK
[4] University of Bergen, Bergen, Norway

Abstract. Lifelogging refers to the process of digitally capturing a continuous and detailed trace of life activities in a passive manner. In order to assist the research community to make progress in the organisation and retrieval of data from lifelog archives, a lifelog task was organised at NTCIR since edition 12. Lifelog-3 was the third running of the lifelog task (at NTCIR-14) and the Lifelog-3 task explored three different lifelog data access related challenges, the search challenge, the annotation challenge and the insights challenge. In this paper we review the dataset created for this activity, activities of participating teams who took part in these challenges and we highlight learnings for the community from the NTCIR-Lifelog challenges.

Keywords: Lifelog · Information retrieval · Test collection

1 Introduction

Lifelogging refers to the process of digitally capturing a detailed trace of life activities in a passive manner [12]. NTCIR-14 hosted the third running of the Lifelog task which aimed to advance efforts at lifelog data analytics and retrieval. Over the three iterations of the task, from NTCIR-12 [9], NTCIR-13 [10] and this year, we report that nearly 20 participating research groups have submitted official runs for the various sub-tasks and we can identify progress in the approaches being made across all tasks, but especially so for the lifelog retrieval task.

Before we begin our review of the submissions for the lifelog task, we introduce the concept of lifelogging by presenting the definition of Dodge and Kitchin [4], who refer to lifelogging as '*a form of pervasive computing, consisting of a unified digital record of the totality of an individual's experiences, captured multimodally through digital sensors and stored permanently as a personal*

M. P. Kato et al. (Eds.): NTCIR 2019, LNCS 11966, pp. 16–28, 2019.
https://doi.org/10.1007/978-3-030-36805-0_2

multimedia archive'. This lifelog task was initially proposed because the organisers identified that technological progress had resulted in lifelogging becoming a more commonplace activity, thereby necessitating the development of new forms of data analytics, organisation and retrieval that are designed to operate over archives of multimodal lifelog data. Additionally, the organisers note recent efforts to employ lifelogging in many domains, such as a means of supporting human memory [13] or facilitating large-scale epidemiological studies in healthcare [25], lifestyle monitoring [27], diet/obesity monitoring [30], or for exploring societal issues such as privacy-related concerns [14] and behaviour analysis [6]. The increasing uptake of lifelogging as a personal and practitioner technology have also lead to related activities in ImageCLEF [2], the Lifelog Search Challenge [11] and a related task at MediaEval 2019. These have all been designed to encourage comparative research into the annotation, retrieval and analysis of multimodal lifelog data.

At NTCIR-14 there were three lifelog sub-tasks, a semantic search sub-task (LEST), a lifelog annotation sub-task (LADT) and an insights sub-task (LIT), of which the LADT was the only new sub-task. In this paper we will provide an overview of the lifelog task, in terms of the dataset, the sub-tasks and the submissions submitted by participating organisations.

2   Task Overview

The Lifelog-3 task explored a number of approaches to information access and retrieval from personal lifelog data, each of which addressed a different challenge for lifelog data organization and retrieval. The three sub-tasks, each of which could have been participated in independently, are as follows:

- **Lifelog Semantic Access sub-Task (LSAT)** to explore search and retrieval from lifelogs.
- **Lifelog Activity Detection sub-Task (LADT)** to identify Activities of Daily Living (ADLs) from lifelogs, which have been employed as indicators of the health of an individual.
- **Lifelog Insight sub-Task (LIT)** to explore knowledge mining and visualisation of lifelogs in an open and topic agnostic manner.

We will now describe and motivate each task in detail.

2.1   LSAT Sub-task

The LSAT sub-task was a known-item search task applied over lifelog data, the aim of which was to advance the state-of-the-art in retrieval of lifelog data in response to typical user information needs. This was created in the spirit of the ad-hoc task at TREC which had lead to the development of many enhanced text search techniques. In this sub-task, the participants had to retrieve a number of specific *moments* in a lifelogger's life in response to a query topic that follows the traditional three-part TREC format of title, description and narrative. We consider moments to be semantic events, or activities that happened at least once in the dataset and are at least one-minute in duration. The task can best be compared to a traditional known-item search task with one (or more) relevant items per topic, though operating over a multimodal dataset. Participants were allowed to undertake the LAST task in an interactive or automatic manner. For interactive submissions, a maximum of five minutes of search time was allowed per topic. The LSAT task included 24 search tasks, generated by the lifeloggers who gathered the data.

2.2   LADT Sub-task

The aim of this sub-task was to develop new approaches to the annotation of multimodal lifelog data in terms of activities of daily living, with a motivation from various healthcare and public-health surveys carried out with lifelog data [25] [30]. An ontology of important lifelog activities of daily living, guided by Kahneman's lifestyle activities [15] were provided as a multi-label classification task. The task required the development of automated approaches for multi-label classification of multimodal lifelog data. Both image content as well as provided metadata and external evidence sources were available to be used to generate the activity annotations. The submission was comprised of one or more activity labels for each image.

2.3   LIT Sub-task

The LIT sub-task was exploratory in nature and the aim of this sub-task was to gain insights into the lifelogger's daily life activities. Participants were requested to provide insights about the lifelog data that support the lifelogger in reflecting upon the data and provide for efficient/effective means of visualisation of the data. There was no explicit evaluation for this topic-agnostic task, so participants were free to analyse and describe the data in whatever manner they wished.

3   Description of the Lifelog-3 Test Collection

As with each of the previous two Lifelog NTCIR tasks, a purpose-built test collection was prepared and released, which was designed to support both ad-hoc retrieval and insights generation from lifelog data. This dataset was prepared following the process described in [3] and took measures to ensure temporal data alignment and privacy preservation of the lifeloggers and bystanders in their data. As with previous tasks, the data was gathered by a number of lifeloggers (in this case, two lifeloggers) who wore various sensing devices and gathered biometric data for most (or all) of the waking hours in the day, along with some manual annotations.

Fig. 1. Examples of wearable camera images from the dataset

3.1 Details of the Dataset

The data consists of a medium-sized collection of rich multimodal lifelog data over 42 days by the two lifeloggers. The contribution of this dataset over previously released lifelog datasets was the inclusion of additional biometric data, a manual diet log and the inclusion of conventional photos. This makes it the richest personal lifelog dataset ever released. The data consists of:

- **Multimedia Content.** Wearable camera PoV images, captured at a rate of two images per minute by an OMG Autographer passive-capture wearable camera, and worn from breakfast to sleep. All recognisable faces and screens were blurred and every image was also resized down to 1024×768 resolution to ensure that text captured was illegible. For examples, see Fig. 1. Additional multimedia content included a time-stamped record of music listening activities sourced from Last.FM[1] and an archive of all conventional (active-capture) digital photos taken by the lifelogger.
- **Biometric Data.** Using FitBit fitness trackers[2], the lifeloggers gathered 24×7 heart rate, calorie burn and steps as numeric lifelog data. In addition, continuous blood glucose monitoring captured readings every 15 min using the Freestyle Libre wearable sensor[3], which provided a continuous record of the blood glucose levels of the individual, which would typically change based on foods consumed and physical activity rate.
- **Human Activity Data.** The daily activities of the lifeloggers were captured in terms of the semantic locations visited, physical activities (e.g. walking, running, standing) from the Moves app[4], along with a time-stamped diet-log

---

[1] Last.FM Music Tracker and Recommender - https://www.last.fm/.

[2] Fitbit Fitness Tracker (FitBit Versa) - https://www.fitbit.com.

[3] Freestyle Libre wearable glucose monitor - https://www.freestylelibre.ie/.

[4] Moves App for Android and iOS - http://www.moves-app.com/.

of all food and drink consumed which was manually recorded by the lifeloggers throughout the day.

- **Enhancements to the Data.** The wearable camera images were annotated with the outputs of a visual concept detector, which provided three types of outputs (attributes, categories and visual concepts). The image attributes and categories of the place in the image are extracted using PlacesCNN [29]. The visual concepts are detected object category and its bounding box extracted by using Faster R-CNN [22] trained on MSCOCO dataset [18]. These formed the official annotations of the visual content of the collection. In some cases, participants added additional non-official annotations using other sources, as outlined below, which in many cases was shown to provide additional beneficial metadata to assist the retrieval and organisation process (Table 1).

Table 1. Statistics of NTCIR-14 lifelog data

	Size
Number of lifeloggers	2
Number of DAYS	43 days
Size of the collection	14 GB
Number of images	81,474 images
Number of locations	61 semantic locations
Number of LSAT topics	24 topics
Number of LADT types	16 activities

3.2 Topics

The LSAT task includes 24 topics with pooled relevance judgements, as is often used in comparative benchmarking activities with large datasets. These LSAT topics were evaluated in terms of traditional Information Retrieval effectiveness measurements such as Precision, RelRet and MAP. These 24 topics were labelled as being one of two types, either precision-based or recall-based. Precision-based topics had a small number of relevant items in the dataset, whereas Recall-based topics would have had a larger number of relevant topics. Each topic was further labelled as being related to User 1, User 2 or both users. An example of a topic is shown in Fig. 2, along with some example relevant image content from the collection. For the full list of the topics, please refer to the NTCIR14-Lifelog3 Overview paper [8] or the associated task website[5].

For the LADT (Activity Detection) sub-task, there were sixteen types of activities defined for annotation, which aim to cover a wide-range of daily human activities. These were defined in order to make it easier for participants to develop event segmentation algorithms for the very subjective human event segmentation

---

[5] http://ntcir-lifelog.computing.dcu.ie/.

TITLE: *Watching Football on TV*
DESCRIPTION: *Find the moments when either U_1 or U_2 was watching football on the TV.*
NARRATIVE: *To be considered relevant, either u1 or u2 must be indoors and watching football on a television. Watching any other TV content is not considered relevant, nor is watching football in a stadium or other external environment.*
EXAMPLES OF RELEVANT IMAGES FOUND BY PARTICIPANTS

Fig. 2. LSAT topic example, including example results.

tasks. The sixteen types of activity were: *traveling, face-to-face interacting, using a computer, cooking, eating, time with children, houseworking, relaxing, reading, socialising, praying, shopping, gaming, physical activities, creative activities, and other activities* (i.e. those not represented by the previous fifteen labels). Each image can be tagged as belonging to one or more activities. A description of each activity is provided in the NTCIR14-Lifelog3 Overview paper [8].

For the LIT task, there were no topics and participants were free to analyse the data in whatever manner they wished in order to extract meaningful insights. One group took part in the LIT task, which is outlined in the relevant section below.

3.3 Relevance Judgement and Scoring

As stated, pooled binary relevance judgements were generated for all 24 LSAT topics. Scoring for the LSAT sub-task was calculated using trec_eval [1]. Two custom applications were developed to support both the LSAT and LADT evaluation processes. For the LADT topics/labels, manual relevance judgements were performed over 5,000 of the images and these annotations were used in assessing participant performance. These images were chosen randomly from the collection and scores were calculated according to the following process. For each run, using the labelled subset of the test images, the score was calculated as the number of correctly predicted labels divided by the total number of labels in the ground truth collection (over all activities). It is worth noting that for some activities, the official runs did not include any labelled images i.e. gaming, praying, physical activity and time with children.

4   Participants and Submissions

In total fourteen participants signed up to the Lifelog-3 task at NTCIR-14, however only five participants submitted to any of the sub-tasks of the Lifelog task. We will now summarise the effort of the participating groups in the sub-tasks that they submitted to.

4.1   LSAT Sub-task

Four participating groups took part in the LSAT sub-task. **NTU (Taiwan)** took part in both the LSAT and LADT Tasks [7]. For the LSAT task, the NTU team developed an interactive lifelog retrieval system that automatically suggested a list of candidate query words to the user, and adopted a probabilistic relevance-based ranking function for retrieval. They enhanced the official concept annotations by applying the Google Cloud Vision API[6] and pre-processed the visual content to remove images with poor quality and to offset the fish-eye nature of the wearable camera data. In the provided examples, this was shown to increase the quality of the non-official annotations. The interactive system facilitated a user to select from suggested query words and to restrict the results to a particular user and date/time interval. Three official runs were submitted, one automatic and two interactive. The first run (NTU-Run1) used an automatic query enhancement process using the top 10 nearest concepts to the query terms. The other two runs employed a user-in-the-loop (NTU-Run2 & NTU-Run3).

QUIK (Japan) from Kyushu University participated in the LSAT task with a retrieval system that integrates online visual WWW content in the search process and operated based on an underlying assumption that a lifelog image of an activity would be similar to images returned from a WWW search engine for similar activities [26]. The approach operated using only the visual content of the collection and used the WWW data to train a visual classifier with a convolutional neural network for each topic. For a given query, images from the WWW were gathered, filtered by a human and combined to create a new visual query (average of 170 images per query). In order to solve the lexical gap between query words and visual concept labels, a second run employed word embedding when calculating the similarities. Two runs were submitted. QUIK-Run1 used only visual concepts while QUIK-Run2 used the visual concepts as well as the query-topic similarity.

VNU-HCM (Vietnam) group took part in the LSAT task by developing an interactive retrieval system [19]. The research required a custom annotation process for lifelog data based on the identifiable habits of the lifeloggers. This operated by extracting additional metadata about each moment in the dataset, by adding in outputs of additional object detectors, manually adding in ten habit concepts, scene classification, and counting the number of people in the images. Associated with this new data source, the team developed a scalable and user-friendly interface that was designed to support novice users to generate queries

---

[6] Google Cloud Vision API - https://cloud.google.com/vision/.

and browse results. One run was submitted (HCMUS-Run1), which was the best performing run at Lifelog-3.

DCU (Ireland) group took part in the LSAT task by developing an interactive retrieval engine for the lifelog data with the official annotations [20]. The retrieval engine was designed to be used by novice users and relied on an extensive range of facet filters for the lifelog data and limited search time to five minutes for each topic. The results of a query were displayed in 5 pages of 20 images, and for any given image, the user could browse the (temporal) context of that image in order to locate relevant content. The user study and subsequent questionnaire illustrated that the interface and search supports provided were generally liked by users.

Table 2. LSAT results for NTCIR-14 Lifelog-3 subtask.

Group ID	Run ID	Approach	MAP	P@10	RelRet
NTU	NTU-Run1	Automatic	0.0632	0.2375	293
NTU	NTU-Run2	Interactive	0.1108	0.3750	464
NTU	NTU-Run3	Interactive	0.1657	0.6833	407
DCU	DCURun1	Interactive	0.0724	0.1917	556
DCU	DCU-Run2	Interactive	0.1274	0.2292	1094
HCMUS	**HCMUS-Run1**	**Interactive**	**0.3993**	**0.7917**	**1444**
QUIK	QUIK-Run1	Automatic	0.0454	0.1958	232
QUIK	QUIK-Run2	Automatic	0.0454	0.1875	232

It can be seen from Table 2 that the results could be analysed by considering both automatic and interactive runs. For automatic runs, NTU achieve the best scores in all three measures: MAP, P@10 and RelRet of 6.32%, 23.75% and 293 respectively while QUIK also generates competitive results. For interactive runs, the team from HCMUS obtains the highest scores of all three measures, which are also the highest results in two approaches with MAP, P@10 and RelRet of 39.93%, 79.17% and 1444 respectively. Whether this performance is due to higher quality annotations or the intuitive interface is not yet clear. While NTU focused on increasing P@10 of their interactive system (68.33%), DCU concentrated on increasing the recall measure by returning as many number of relevant images as possible (RelRet: 1094 images). Both teams managed to achieve the second highest scores of the corresponding measure system. We can hypothesise that these findings suggest that enhancing the official visual annotations with higher-quality non-official annotations seems to lead to an enhanced performance in both automatic and interactive retrieval. This finding is in line with similar findings from previous editions of the lifelog task at NTCIR-12 [9] and NTCIR-13 [10], as well as findings from the related LSC search challenge [11].

4.2   LADT Task

The NTU group (Taiwan) took part in the LADT task [7] and developed a new approach for the multi-label classification of lifelog images. In order to train the classifier, the authors manually labelled four days, which were chosen because they covered most of the activities that the lifeloggers were involved in. It is noted that there is no training data generated for some of the activities for user 1 and user 2. Since only one group took part, no comparison is possible between participants. Readers are referred to the NTU paper [7] for details of their different runs and the comparative performance of these runs.

4.3   LIT Sub-task

For the LIT task, there were no submissions to be evaluated in the traditional manner; rather the LIT task was an exploratory task to explore a wide-range of options for generating insights from the lifelog data. One group took part in the LIT task. **THUIR (China)** developed a number of detectors for the lifelog data to automatically identify the status/context of a user [19], which could be used in many real-world applications, especially so for forms of assistive technology. There were three detectors developed for inside/outside status, alone/not alone status and working/not working status. These detectors were designed to operate over booth non-visual and visual lifelog data. A comparison between the two approaches showed that the visual features (integrating supervised machine learning) were significantly better than non-visual ones based on metadata. Finally the authors presented a number of statistics of users' activities for all three detectors, which highlighted the activities of the two users in a highly visual manner.

5   Advancing State-of-the-art over Three Editions

Lifelog-3 was the third in a series of collaborative benchmarking exercises for lifelog data at NTCIR. It attracted five active participants, four for the automatic LSAT sub-task, one for the LADT sub-task and one for the LIT sub-task. Over the course of the three editions, we can identify a number of successful techniques that can be applied over lifelog data to enhance the performance of both automatic and interactive systems. We can summarise the advances in state-of-the-art as follows:

- The utilisation of **non-official visual concept detectors** is considered a positive addition. The official visual concepts released with the test collections were using off-the-shelf approaches, but customised techniques or employing latest approaches from the field were shown to significantly advance performance. For example, at NTCIR-12, the LIG-MRM group (France) performed significantly ahead of all other submissions, by focusing on enhancing the performance of the visual concept detectors to be used for retrieval [23]. Likewise the VCI2R at NTCIR-13 proposed a general framework to bridge the semantic gap between lifelog data and the event-based LSAT topics [17] by enhancing the visual annotations and employing temporal smoothing of annotations, which proved to be the most successful approach at NTCIR-13. At NTCIR-14, the VNU-HCM (Vietnam) group developed an interactive retrieval system [19] that used enhanced visual metadata (including human annotations), which outperformed all other approaches.

– We note the integration of **additional metadata sources** in some approaches, which was also considered to be beneficial. For example, at NTCIR-12 the QUIK team (Japan) integrated online visual WWW content in the search process to enhance the performance of image-based retrieval by using the WWW data to train a visual classifier with a convolutional neural network for each topic [26]. Also at NTCIR-12, the VTIR team identified that location was a very important component in the information retrieval process [28] and thus enhanced location semantic descriptions were used to facilitate retrieval.

– Another source of enhancement was taking measures to **reduce the lexical gap** between user queries and concept annotations using some form of term expansion, and the current consideration is that this could be achieved using word embedding approaches. At NTCIR-12, the QUT group took an approach to retrieval that utilised spreading annotations (via visual similarity) to generate long, descriptive paragraphs of text to annotate the lifelog content, as opposed to the conventional tag-based approach [24]. The IDEAS Institute for Information Industry (Taiwan) took a textual approach to retrieval [16] at NTCIR-14, utilising word2vec to better match visual concepts to user queries (an approach referred to as bridging the lexical gap at NTCIR-14) via query expansion.

– We note the increasing use of **interactive systems** throughout the editions to address the challenges posed by the LSAT task. At NTCIR-12, there was one interactive retrieval system presented by the team from the University of Barcelona who developed an interactive retrieval engine that integrated a semantic-content tagging tool to enhance the quality of the annotations [21], which naturally outperformed all automatic approaches. At NTCIR-13, the DCU team employed a human-in-the-loop to translate the provided queries into system queries for their retrieval engine, in one of their runs [5]. However, at NTCIR-14, we note that three of the participants developed interactive systems and a fourth participant also integrated the human-in-the-loop query enhancement. This interest in interactive retrieval systems has lead to the development of a parallel benchmarking activity, the Lifelog Search Challenge for interactive retrieval systems [11].

At this point, after the three editions of the lifelog task, the main approaches that the organisers' consider to be valuable for lifelog access is the use of enhanced visual concept detectors to improve indexing and the application of approaches to bridging the lexical gap, either via some form of index term expansion or query-expansion.

6   Conclusion

In this paper, we described the data and the activities from the Lifelog-3 core-task at NTCIR-14, which was the third and final edition of the NTCIR-Lifelog task, which included three sub-tasks, each of which addressed a different challenge for lifelog organisation and retrieval. For the LSAT sub-task, four groups took part and produced eight official runs including five interactive and three automatic runs. The approach taken by HCMUS, of enhancing the provided annotations with additional object detectors, habits, scenes and people analytics, along with an intuitive user interface, ensured that their runs were significantly better than the runs of any other participant. The LADT and LIT tasks attracted one participant each, so we are not in a position to draw any conclusions at this point.

After this, the third instance of the NTCIR-Lifelog task, we are beginning to see some learnings from the comparative benchmarking exercises. It can be seen that additional visual concept detectors, integrating external sources and addressing the lexical gap between users and the systems are priority topics for the research community to address. Likewise we note the interest in the community of developing interactive (user-in-the-loop) approaches to lifelog data retrieval. What we have not yet seen is widespread use of the temporally aligned metadata that accompanies the lifelog datasets. We hope that participants and readers will continue the effort to develop new approaches for the organisation and retrieval of lifelog data, and take part in future NTCIR, LSC and Image-CLEF efforts within the domain.

Acknowledgements. This publication has emanated from research supported in party by research grants from Irish Research Council (IRC) under Grant Number GOIPG/2016/741 and Science Foundation Ireland under grant numbers SFI/12/RC/2289 and SFI/13/RC/2106. We acknowledge the support and input of the DCU ethics committee and the risk & compliance officer.

References

1. Buckley, C.: Treceval IR evaluation package (2004)
2. Dang-Nguyen, D.T., Piras, L., Riegler, M., Zhou, L., Lux, M., Gurrin, C.: Overview of ImageCLEFlifelog 2018: daily living understanding and lifelog moment retrieval. In: CLEF2018 Working Notes. CEUR Workshop Proceedings, 10–14 September 2018. CEURWS, Avignon (2018)
3. Dang-Nguyen, D.T., Zhou, L., Gupta, R., Riegler, M., Gurrin, C.: Building a disclosed lifelog dataset: challenges, principles and processes. In: Content-Based Multimedia Indexing (CBMI) (2017)
4. Dodge, M., Kitchin, R.: 'Outlines of a world coming into existence': pervasive computing and the ethics of forgetting. Environ. Plan. **34**(3), 431–445 (2007). https://doi.org/10.1068/b32041t
5. Duane, A., Zhou, L., Dang-Nguyen, D.T., Gurrin, C.: DCU at the NTCIR-13 lifelog-2 task. In: Proceedings of NTCIR-13, Tokyo, Japan (2017)

6. Everson, B., Mackintosh, K.A., McNarry, M.A., Todd, C., Stratton, G.: Can wearable cameras be used to validate school-aged children's lifestyle behaviours? Children 6(2), 20 (2019). https://doi.org/10.3390/children6020020

7. Fu, M.H., Chia-Chun, C., Huang, G.H., Chen, H.H.: Introducing external textual knowledge for lifelog retrieval and annotation. In: The Fourteenth NTCIR Conference (NTCIR-14) (2019)

8. Gurrin, C., et al.: Overview of the NTCIR-14 lifelog-3 task. In: The Fourteenth NTCIR Conference (NTCIR-14) (06 2019)

9. Gurrin, C., Joho, H., Hopfgartner, F., Zhou, L., Albatal, R.: Overview of NTCIR-12 lifelog task. In: Kando, N., Kishida, K., Kato, M.P., Yamamoto, S. (eds.) Proceedings of the 12th NTCIR Conference on Evaluation of Information Access Technologies, pp. 354–360 (2016)

10. Gurrin, C., et al.: Overview of NTCIR-13 lifelog-2 task. In: The Thirteenth NTCIR conference (NTCIR-13), pp. 6–11 (2017)

11. Gurrin, C., et al.: Comparing approaches to interactive lifelog search at the lifelog search challenge (LSC2018). ITE Trans. Media Technol. Appl. 7(2), 46–59 (2019)

12. Gurrin, C., Smeaton, A.F., Doherty, A.R.: LifeLogging: personal big data. Found. Trends® Inform. Retrieval 8(1), 1–125 (2014). https://doi.org/10.1561/1500000033. http://www.nowpublishers.com/articles/foundations-and-trends-in-information-retrieval/INR-033

13. Harvey, M., Langheinrich, M., Ward, G.: Remembering through lifelogging: a survey of human memory augmentation. Pervasive Mob. Comput. 27, 14–26 (2016). https://doi.org/10.1016/j.pmcj.2015.12.002

14. Hoyle, R., Templeman, R., Armes, S., Anthony, D., Crandall, D., Kapadia, A.: Privacy behaviors of lifeloggers using wearable cameras. In: Proceedings of the 2014 ACM International Joint Conference on Pervasive and Ubiquitous Computing, UbiComp 2014, pp. 571–582. ACM, New York (2014)

15. Kahneman, D., Krueger, A.B., Schkade, D.A., Schwarz, N., Stone, A.A.: A survey method for characterizing daily life experience: the day reconstruction method. Science 306, 1776–1780 (2004)

16. Lin, H.L., Chiang, T.C., Chen, L.P., Yang, P.C.: Image searching by events with deep learning for NTCIR-12 lifelog. In: Proceedings of NTCIR-12, Tokyo, Japan (2016)

17. Lin, J., del Molino, A.G., Xu, Q., Fang, F., Subbaraju, V., Lim, J.H.: VCI2R at the NTCIR-13 lifelog semantic access task. In: Proceedings of NTCIR-13, Tokyo, Japan (2017)

18. Lin, T., et al.:: Microsoft COCO: common objects in context. CoRR abs/1405.0312 (2014). http://arxiv.org/abs/1405.0312

19. Nguyen, I.V.K., Shrestha, P., Zhang, M., Liu, Y., Ma, S.: THUIR at the NTCIR-14 lifelog-3 task: how does lifelog help the user's status recognition. In: The Fourteenth NTCIR Conference (NTCIR-14) (2019)

20. Ninh, V.T., et al.: A baseline interactive retrieval engine for the NTICR-14 lifelog-3 semantic access task. In: The Fourteenth NTCIR Conference (NTCIR-14) (2019)

21. de Oliveira Barra, G., et al.: LEMoRe: a lifelog engine for moments retrieval at the NTCIR-lifelog lsat task. In: Proceedings of NTCIR-12, Tokyo, Japan (2016)

22. Ren, S., He, K., Girshick, R.B., Sun, J.: Faster R-CNN: towards real-time object detection with region proposal networks. CoRR abs/1506.01497 (2015). http://arxiv.org/abs/1506.01497

23. Safadi, B., Mulhem, P., Qul'not, G., Chevallet, J.P.: MRIM-LIG at NTCIR lifelog semantic access task. In: Proceedings of NTCIR-12, Tokyo, Japan (2016)

24. Scells, H., Zuccon, G., Kitto, K.: QUT at the NTCIR lifelog semantic access task. In: Proceedings of NTCIR-12, Tokyo, Japan (2016)
25. Signal, L.N., et al.: KidsCam: an objective methodology to study the world in which children live. Am. J. Prevent. Med. **53**(3), e89–e95 (2017). https://doi.org/10.1016/j.amepre.2017.02.016
26. Suzuki, T., Ikeda, D.: Smart lifelog retrieval system with habit-based concepts and moment visualization. In: The Fourteenth NTCIR Conference (NTCIR-14) (2019)
27. Wilson, G., Jones, D., Schofield, P., Martin, D.J.: The use of a wearable camera to explore daily functioning of older adults living with persistent pain: methodological reflections and recommendations. J. Rehabil. Assist. Technol. Eng. **5**, 2055668318765411 (2018). https://doi.org/10.1177/2055668318765411
28. Xia, L., Ma, Y., Fan, W.: VTIR at the NTCIR-12 2016 lifelog semantic access task. In: Proceedings of NTCIR-12, Tokyo, Japan (2016)
29. Zhou, B., Lapedriza, A., Khosla, A., Oliva, A., Torralba, A.: Places: a 10million image database for scene recognition. IEEE Trans. Pattern Anal. Mach. Intell. **40**, 1452–1464 (2017)
30. Zhou, Q., et al.: The use of wearable cameras in assessing children's dietary intake and behaviours in China. Appetite **139**, 1–7 (2019)

A Baseline Interactive Retrieval Engine for Visual Lifelogs at the NTCIR-14 Lifelog-3 Task

Van-Tu Ninh[1], Tu-Khiem Le[1], Liting Zhou[1], Graham Healy[1],
Kaushik Venkataraman[1], Minh-Triet Tran[2], Duc-Tien Dang-Nguyen[3],
Sinead Smyth[1], and Cathal Gurrin[1(✉)]

[1] Dublin City University, Dublin, Ireland
`cathal.gurrin@dcu.ie`
[2] VNU-HCM, University of Science, Ho Chi Minh City, Vietnam
[3] University of Bergen, Bergen, Norway

Abstract. This paper describes the work of DCU research team in collaboration with University of Science, Vietnam, and University of Bergen, Norway at the Lifelog task of NTCIR-14. In this paper, a new interactive retrieval engine is described that supports faceted retrieval and we present the results of an initial experiment with four users. Following this initial experiment, we implement a list of changes for a revised interactive retrieval engine for the LSC2019 comparative evaluation competition. The interactive retrieval system we describe utilises the wide range of lifelog metadata provided by the task organisers to develop an extensive faceted retrieval system.

Keywords: Interactive lifelog search engine · Information retrieval

1 Introduction

Information Retrieval has a long history of utilising the human as a key component of a retrieval system. Our current generation of WWW search engines rely on the human as an integral part of the search process, in terms of query generation, refinement and result selection. Inspired by the 'human-in-the-loop' model of interactive information retrieval, the DCU team, with the support of VNU-HCM, University of Science and the University of Bergen, developed a prototype interactive retrieval system for the LSAT - Lifelog Semantic Access subtask of the NTCIR-14 Lifelog task [3]. In this paper we introduce this prototype retrieval engine, we present the performance of the retrieval engine in the LSAT task, we report on the findings of a small-scale qualitative user study of the prototype, and we highlight the enhancements carried out on the prototype for our participation in the LSC2019 Lifelog Search Challenge.

V.-T. Ninh and T.-K. Le—The two authors contributed equally to this paper.

© Springer Nature Switzerland AG 2019
M. P. Kato et al. (Eds.): NTCIR 2019, LNCS 11966, pp. 29–41, 2019.
https://doi.org/10.1007/978-3-030-36805-0_3

2   Related Interactive Lifelog Retrieval Systems

The Lifelog Semantic Access Task, which began in NTCIR-12, allows both automatic and interactive lifelog search systems [4] to be comparatively evaluated in an open benchmarking exercise. In NTCIR-12, the University of Barcelona and Technical University of Catalonia developed an interactive search engine [16] which utilised visual semantic concepts from images and used them as tags for the interactive image retrieval system. They also employed WordNet to create the similarity between tags to assist novice/expert users to choose the most relevant appropriate tags. Moreover, a heatmap was generated to show the confidence of the retrieval result which aims to achieve the best configuration of precision and recall of their retrieval system. In the official results of the lifelog task, their best run (unsurprisingly) outperforms all the best ones of other teams that built automatic search engines [4]. For the LSAT task at NTCIR-14, [2] developed an interactive lifelog retrieval system that automatically suggested a list of candidate query words to the user and adopted a probabilistic relevance-based ranking function for retrieval. They enhanced the official concept annotations by applying the Google Cloud Vision API[1] and pre-processed the visual content to remove images with poor quality and to offset the fish-eye nature of the wearable camera data. In the provided examples, this was shown to increase the quality of the non-official annotations. Additionally, an interactive system was developed by [15], which operated as a faceted search system over enhanced metadata (additional object detectors, manually adding in ten habit concepts, scene classification, and counting the number of people in the images). The user interface was designed to be user-friendly and support novice users to generate queries and browse results. The system performed significantly better than any other interactive system at NTCIR-14, including the system described in this paper.

More recently, we note the introduction of a new challenge, specifically aimed at comparing approaches to interactive retrieval from lifelog archives. The Lifelog Search Challenge (LSC) [6] utilises a similar dataset [5] to the one used for the NTCIR14-Lifelog task. The LSC has occurred in 2018 and 2019 and attracted significant interest from participants. We report on the some of the most relevant of these here. In 2018, six participating teams took part in the live search challenge. These teams had all indexed the dataset prior to attending the workshop and then during the interactive search challenge, both expert and novice users took part in evaluating the performance of the six systems.

Alpen-Adria-Universität Klagenfurt (AAU) developed an interactive retrieval engine based on a video-retrieval system. Called liveXplore [14], a system modification serving as a lifelogging data browser by focusing on visual exploration and retrieval as well as metadata filtering. The system focused on visual similarity, concept and metadata filtering; it performed very well in 2018, coming a very close second place to the eventual winner [1]. A similar version of liveXplore was deployed for the LSC2019 challenge [10], though not as successfully.

---

[1] Google Cloud Vision API - https://cloud.google.com/vision/.

Another system of note came from Charles University, Prague, with a repurposed an updated version of the VIRET video retrieval system [12]. Every day from the collection was treated as one 'video' represented by the lifelog images, with automatic annotations associated with each image using GoogleNet. In addition, a colour signature for sketch-based search and deep feature vector from the original GoogleNet were extracted. The system came a close third place in both the 2018 and 2019 completions. Additionally in 2019, we note two additional systems that warrant review. The vitrivr system from Rossetto et al. [19] was an enhanced version of the vitrivr open-source multimedia retrieval system, which was developed for video retieval tasks. Extensions to the leading interactive video retrieval system included the capability to process Boolean query expressions alongside content-based query descriptions in order to leverage the structural diversity inherent to lifelog data. This system was the eventual winner of the LSC 2019 competition. A final system of note was developed by [8] which, as per their work at NTCIR-14 [15] included additional enhanced metadata that proved meaningful for the retrieving process, and a user interface that was designed to support a novice user to perform the retrieval efficiently.

3   An Interactive Lifelog Retrieval System

For the LSAT sub-task, we developed a retrieval system to provide timely, precise and convenient access to a lifelog data archive. The system, as well as our official submissions were designed to maximise recall, in order to support a user to access their life experiences in a real-world lifelogging scenario.

3.1   Data Preprocessing

After analyzing NTCIR-14 lifelog data [3], we divide the data into five main categories: time, location, activities, biometrics and visual concepts.

1. **Time:** For time data, we split the minute based lifelog data into selection of range of hours/minutes/days/day of week for lifelog search engine. Novice/expert user can utilise this information to narrow the scope of searching for a topic in lifelogger's data. All time data is converted into the UTC time standard.
2. **Location:** For location data, we also utilise timezone information to know the region/country where the lifelogger is visiting. We convert locations into semantic names to help novice/expert user locate the category of place when searching for lifelogger's moments.
3. **Activity:** The activity data contains two categories: walking and transport.
4. **Biometrics:** The biometrics data that we use in our search engine includes heart rate and calories.
5. **Visual concepts:** We included the visual concepts that were distributed with the dataset. Visual concepts are of three types: place attribute, place category, and visual objects. The place features were extracted using places365-CNN [20]. The visual objects' categories originate from MSCOCO dataset [11] and are automatically detected using object detection network [17].

3.2   Supporting Faceted Search

The interactive retrieval engine implemented a faceted search system in which a user could either enter a textual query in a conventional text box, or select from a range of facets of the metadata to locate items of interest. The faceted search system operated over a range of metadata which are listed in Subsect. 3.1 which are day of the week, date, time range, user activity (walking/transporting), biometrics data ranges (calories and heart rate), location (location category and name), and visual concepts (place attributes, place categories, and detected objects) in the corresponding order.

When searching using the conventional text box, a user is limited to utilising only visual concepts, activity, and location, and as such, it was a simplified version of a conventional bag-of-words retrieval system. If user desired to utilise all the metadata in searching for relevant items, the faceted query mechanism was required to support this.

The interface, showing the faceted panel (left), the querybox (top) and the result browsing panel (right) is shown in Fig. 1. Note the timer on the top of the main panel, which was added to support the LSAT interactive experiment.

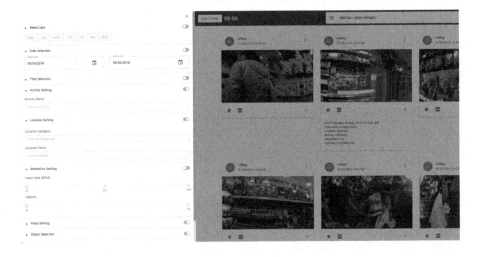

Fig. 1. Query panel

Upon generating a query, the system generated a list of results (20 per page and 5 pages of results) ranked in temporal order, as shown in Fig. 2, using a conventional text ranking algorithm. The unit of retrieval was the image, as was expected for the LSAT task. Each image is given a title, date, a button to select the image and another one to show before & after the current image. Summary metadata from each image could be displayed by selecting the image. If an image was selected as being relevant (the star icon), then it was saved for submission.

Fig. 2. Result panel

Submission occurred automatically after a given time period had elapsed, in our case, this was five minutes.

Additionally, for any image, the temporal context was made available by selecting the double box under each image. The temporal context appeared as a hovering panel and the user could browse back (left) or forward (right) in time, see Fig. 3 for the temporal context of an image for the topic 'toystore'. Selecting an image allowed it to be flagged as relevant.

At the end of a five minute period, all saved images were used to form the official submission. Additionally, all images immediately before and after (to a depth of ten) were appended to the end of the official submission for evaluation. The idea was that additional relevant content could be found in the temporal neighbourhood of every relevant image. The rank order of submissions was in the order that the user selected the relevant items, followed by the temporal neighbourhood images. In this way, the system maximised the potential for Recall, though at the expense of measures such as MAP.

4   Interactive Experiment

In order to submit the official runs for the NTCIR14-Lifelog3 LSAT subtask, we organised an interactive user experiment in which novice users used the interactive retrieval system according to the following parameters and protocol.

4.1   Experimental Configuration

The evaluation was performed by four novice users whom each executed twelve topics. The topics were divided into two groups (1 → 12 and 13 → 24). Each user was given five minutes to complete each topic. The experimental protocol was as

Fig. 3. Temporal browsing

follows. The participant was introduced to the system and given a five minute review of functionality. Following this, the participant was allowed to test the system for a further ten minutes with two sample queries. Once the participant was comfortable with the system and how it operated, the user study began with the user reading a topic and the five minute timer was started once the user was comfortable that they understood the topic. All twelve topics were executed in forward order for users 1 and 2, and in reverse order for users 3 and 4. It would take around 90 min per user to conduct the whole experiment. In terms of practical experimental configuration, two users took part in the experiment in parallel (1 and 2, followed by 3 and 4).

4.2   Results

The user experiment produced two runs; one combining the submissions of DCU-run1 (users 1, 2), and a second for DCU-run2(users 3, 4). DCU-run1 contained submitted results for 22 of the 24 topics, whereas DCU-run2 contained results for 23 of the 24 topics. For missing topics, it means that the user could not find any relevant images which are suitable to the detailed description of the topic. The total number of retrieved relevant results for DCU-run1 was 556, whereas for DCU-run2, it was 1094. DCU-run2 users found significantly more results that DCU-run1, which highlights a variability in how the teams were formed. Interestingly, users 3 & 4 would have scored the system usability lower than users 1 & 2, although their interaction found over double the number of relevant items.

Considering that we were employing pagination of results at 20 per page, the P@10 metric for DCU-run1 was 0.1917 but for DCU-run2, it was 0.2292. Given the nature of the experiment, exploring results from a ranked list at higher cut-

off points was not valuable due to less similarity to the query's content. In terms
of MAP, DCU-run1 was 0.0724, but for DCU-run2 it was 0.1274, once again
significantly higher.

When comparing performance of this system with other participants' interactive retrieval system in the LSAT sub-task (see Table 1), it is apparent that
the DCU-Run1 underperformed against other runs in terms of MAP and P@10,
with only DCU-Run2 performing better than any competitor. It is our conjec-
ture that this was due to the packing of the result submission with the temporal
images, which would have reduced the MAP and P@10 scores. Considering the
RelRet (Relevant items Recalled) measure, both runs were only bettered by the
HCMUS system [15], which was the overall best performing interactive system.
Once again, the submission packing would have increased these RelRet scores.
Another factor that could be taken into consideration was the application of a
five-minute time limit on each topic. Had this been longer, then the scores would
likely have changed.

Table 1. Comparing DCU-Run1 & 2 with other Interactive Runs, from [3]

Group ID	Run ID	Approach	MAP	P@10	RelRet
NTU	NTU-Run2	Interactive	0.1108	0.3750	464
NTU	NTU-Run3	Interactive	0.1657	0.6833	407
HCMUS	HCMUS-Run1	Interactive	0.3993	0.7917	1444
DCU	**DCU-Run1**	**Interactive**	**0.0724**	**0.1917**	**556**
DCU	**DCU-Run2**	**Interactive**	**0.1274**	**0.2292**	**1094**

When considering both automatic and interactive retrieval efforts, the best
run that automatic retrieval system achieved was from the NTU group (MAP
= 0.0632, P@10 = 0.2375, RelRet = 293) [2]. Although their P@10 is slightly
higher than our baseline interactive retrieval system, its both RelRet and MAP
are not as good as our baseline system. However, the high P@10 score gives
promising improvement on automatic retrieval search engine. The overall result
of NTCIR-14 for both interactive and automatic retrieval manner could be seen
in Table 2.

4.3 User Feedback

The inter-run comparisons just presented are not very useful when considering
how well a system is liked by users. Clearly users 3 and 4 outperformed users 1
and 2. Using a questionnaire (The User Experience Questionnaire - QEU) [7],
we sought to get an initial feedback from users about their experiences with the
interactive retrieval engine. All four users filled in the simple 8 part questionnaire,
which evaluated the system in terms of pragmatic (realistic-use-case) quality
and hedonic (pleasantness) quality, with results shown in Table 3. In terms of

Table 2. Comparing DCU-Run1 & 2 with Automatic Runs, from [3]

Group ID	Run ID	Approach	MAP	P@10	RelRet
NTU	NTU-Run1	Automatic	0.0632	0.2375	293
NTU	NTU-Run2	Interactive	0.1108	0.3750	464
NTU	NTU-Run3	Interactive	0.1657	0.6833	407
HCMUS	HCMUS-Run1	Interactive	0.3993	0.7917	1444
QUIK	QUIK-Run1	Automatic	0.0454	0.1958	232
QUIK	QUIK-Run2	Automatic	0.0454	0.1875	232
DCU	**DCU-Run1**	**Interactive**	**0.0724**	**0.1917**	**556**
DCU	**DCU-Run2**	**Interactive**	**0.1274**	**0.2292**	**1094**

pragmatic quality, the interface was seen as being slightly more (+0.5 from a maximum of 3.0) supportive than obstructive, slightly more easy (+0.3) than complicated and slightly more clear (+0.3) than confusing. However users felt that it was slightly more inefficient (−0.3) than efficient. In terms of hedonic quality the interface was considered to be significantly more exciting (+1.3) than boring, significantly more interesting (+2.0) than non-interesting, significantly more inventive (+1.3) than conventional and slightly more leading-edge than usual/conventional.

Table 3. Pragmatic quality feedback of DCU's interactive retrieval engine

Item	Mean	Variance	Std. Dev.	Negative	Positive	Scale
1	0.5	6.3	2.5	Obstructive	Supportive	Pragmatic quality
2	0.3	7.6	2.8	Complicated	Easy	
3	−0.3	2.9	1.7	Inefficient	Efficient	
4	0.3	2.9	1.7	Confusing	Clear	
5	1.3	4.3	2.1	Boring	Exciting	Hedonic quality
6	2.0	1.3	1.2	Not interesting	Interesting	
7	1.3	2.3	1.5	Conventional	Inventive	
8	0.8	2.3	1.5	Usual	Leading edge	

Exploring the qualitative findings on a per run basis, DCU-run1 users considered that the system was more supportive, easier to user, more efficient and clearer than DCU-run2 users. In terms of hedonic quality, they also found it more exciting, interesting, inventive and leading edge. However, considering the actual runs, these users were significantly less effective when using the system.

This feedback is reasonable because DCU-run2 users have prior experience of developing application system, which is why they expect the search engine to be more effective, clearer, and less complicated in interacting with our system. In contrast, DCU-run1 users understand how our search engine work after training without any further expectation of user interaction and think that our available functions are enough to retrieve the correct moments.

Through feedback and observation of the users using the retrieval system, we gathered findings that are being used to improve the current system for the LSC'19 (Lifelog Search Challenge) comparative benchmarking exercise. The new system called LifeSeeker [9] is an evolution of this system that incorporated the following updates:

- Taking measures to reduce the lexical gap (between user queries and the indexed concepts within the system) by expanding the indexed terms to include synonyms. Hence, we enriched the output of the visual and biometric concept detectors using a term-expansion (thesaurus-lookup) approach. For example the concept *seaside* would include the following synonyms; *shore, coast, sands, margin, strand, seaside, shingle, lakeside, water's edge, lido, foreshore, seashore, plage, littoral, sea.*
- Integrating content-similarity to allow the user to find similar looking content for any given image. Such a feature had been used successfully by participants in LSC'18. For this we utilised the Bag-of-Words model to transform visual features into a vector representation for comparing and returning similar images. Extracting visual features from image was done thanks to the Scale-Invariant-Feature-Transform (SIFT) [13] detector and cosine vector distance as a dissimilarity measure.
- Including a more conventional free-text search element and integrating the filter panel as part of the free-text query mechanism. The free-text ranking engine implemented in the system indexed all textual content associated with any image within the collection. In order to reduce the architectural complexity and latency of the system, we choose to use a standard approach to term weighting [18]. For the purposes of this interactive system, both stemming and stop-words were employed. The maximum number of results returned was 1,000, although in a standard configuration, only 100 were displayed to the user in the interface. The top 1,000 images was necessary for the ranking system to support faceted filtering.

These changes were combined with a slightly revised interface to take in to account the richer metadata and the content similarity functionality, as shown in Figs. 4, 5, and 6.

In the interactive search competition at LSC2019, this system performed among the top-ranked teams with an overall score of 68, compared to the vitrivr system [19] which was given a score of 100. Interestingly the system significantly closed the gap to the NTCIR-14 system from HCMUS (which also competed at the LSC in 2019) who scored 72 in the competition. Details of the scoring function employed can be found in the review of the LSC 2018 competition [6].

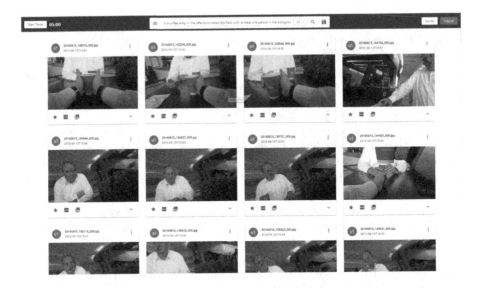

Fig. 4. LifeSeeker interface with free-text search

Fig. 5. LifeSeeker interface with visual content similarity function. The items in the grid belong to the image inside the red bounding box (Color figure online)

Fig. 6. LifeSeeker interface with additional facet filters

5 Conclusions and Future Work

In this paper, we introduced a first-generation prototype of an interactive retrieval engine for lifelog data, that was run at the NTCIR14-Lifelog3 task and enhanced to be competitive in the second LSC Challenge in 2019. The system was a baseline retrieval system that operated over the provided metadata for the collection. The system was evaluated by four users and findings indicate that the system can be effectively used to locate relevant content. User studies showed that the users generally liked the system, but both observation and feedback provided a list of proposed enhancements to the system, which have been integrated into a new interactive retrieval system called LifeSeeker [9] which was shown to be among the best performers at the LSC2019.

Acknowledgements. This publication has emanated from research supported in part by research grants from Irish Research Council (IRC) under Grant Number GOIPG/2016/741 and Science Foundation Ireland under grant numbers SFI/12/RC/2289 and 13/RC/2106. We wish to also thank the four users who used our retrieval system to generate the official runs.

References

1. Duane, A., Gurrin, C., Hürst, W.: Virtual reality lifelog explorer: lifelog search challenge at ACM ICMR 2018. In: Proceedings of the 2018 ACM Workshop on The Lifelog Search Challenge, LSC 2018, pp. 20–23. ACM, New York (2018). https://doi.org/10.1145/3210539.3210544

2. Fu, M.H., Chia-Chun, C., Huang, G.H., Chen, H.H.: Introducing external textual knowledge for lifelog retrieval and annotation. In: The Fourteenth NTCIR conference (NTCIR-14) (2019)

3. Gurrin, C., et al.: Overview of NTCIR-14 lifelog task. In: Proceedings of the 14th NTCIR Conference on Evaluation of Information Access Technologies, National Center of Sciences, 10–13 June 2019, Tokyo, Japan (2019)

4. Gurrin, C., Joho, H., Hopfgartner, F., Zhou, L., Albatal, R.: Overview of NTCIR-12 lifelog task. In: Proceedings of the 12th NTCIR Conference on Evaluation of Information Access Technologies, National Center of Sciences, 7–10 June 2016, Tokyo, Japan (2016)

5. Gurrin, C., et al.: A test collection for interactive lifelog retrieval. In: Kompatsiaris, I., Huet, B., Mezaris, V., Gurrin, C., Cheng, W.-H., Vrochidis, S. (eds.) MMM 2019. LNCS, vol. 11295, pp. 312–324. Springer, Cham (2019). https://doi.org/10.1007/978-3-030-05710-7_26

6. Gurrin, C., et al.: Comparing approaches to interactive lifelog search at the lifelog search challenge (LSC2018). ITE Trans. Media Technol. Appl. **7**(2), 46–59 (2019)

7. Laugwitz, B., Held, T., Schrepp, M.: Construction and evaluation of a user experience questionnaire. In: Holzinger, A. (ed.) USAB 2008. LNCS, vol. 5298, pp. 63–76. Springer, Heidelberg (2008). https://doi.org/10.1007/978-3-540-89350-9_6

8. Le, N.K., et al.: Smart lifelog retrieval system with habit-based concepts and moment visualization. In: Proceedings of the ACM Workshop on Lifelog Search Challenge, LSC 2019, pp. 1–6. ACM, New York (2019). https://doi.org/10.1145/3326460.3329155

9. Le, T.K., et al.: LifeSeeker - interactive lifelog search engine at LSC 2019. In: Proceedings of the 2019 ACM Workshop on The Lifelog Search Challenge, LSC 2019. ACM, New York (2019)

10. Leibetseder, A., et al.: lifeXplore at the lifelog search challenge 2019. In: Proceedings of the ACM Workshop on Lifelog Search Challenge, LSC 2019, pp. 13–17. ACM, New York (2019). https://doi.org/10.1145/3326460.3329157

11. Lin, T.-Y., et al.: Microsoft COCO: common objects in context. In: Fleet, D., Pajdla, T., Schiele, B., Tuytelaars, T. (eds.) ECCV 2014. LNCS, vol. 8693, pp. 740–755. Springer, Cham (2014). https://doi.org/10.1007/978-3-319-10602-1_48

12. Lokoč, J., Souček, T., Kovalčik, G.: Using an interactive video retrieval tool for lifelog data. In: Proceedings of the 2018 ACM Workshop on The Lifelog Search Challenge, LSC 2018, pp. 15–19. ACM, New York (2018). https://doi.org/10.1145/3210539.3210543

13. Lowe, D.G.: Distinctive image features from scale-invariant keypoints. Int. J. Comput. Vis. **60**(2), 91–110 (2004). https://doi.org/10.1023/B:VISI.0000029664.99615.94

14. Münzer, B., Leibetseder, A., Kletz, S., Primus, M.J., Schoeffmann, K.: lifeXplore at the lifelog search challenge 2018. In: Proceedings of the 2018 ACM Workshop on The Lifelog Search Challenge, LSC 2018, pp. 3–8. ACM, New York (2018). https://doi.org/10.1145/3210539.3210541

15. Nguyen, I.V.K., Shrestha, P., Zhang, M., Liu, Y., Ma, S.: THUIR at the NTCIR-14 lifelog-3 task: how does lifelog help the user's status recognition. In: The Fourteenth NTCIR Conference (NTCIR-14) (2019)
16. de Oliveira Barra, G., Ayala, A.C., Bolaños, M., Dimiccoli, M., Giró i Nieto, X., Radeva, P.: LEMoRe: a lifelog engine for moments retrieval at the NTCIR-lifelog LSAT task. In: Proceedings of the 12th NTCIR Conference on Evaluation of Information Access Technologies, National Center of Sciences, 7–10 June 2016, Tokyo, Japan (2016)
17. Ren, S., He, K., Girshick, R., Sun, J.: Faster R-CNN: towards real-time object detection with region proposal networks. In: Proceedings of the 28th International Conference on Neural Information Processing Systems, NIPS 2015, vol. 1, pp. 91–99. MIT Press, Cambridge (2015). http://dl.acm.org/citation.cfm?id=2969239.2969250
18. Robertson, S.E., Jones, K.S.: Simple, proven approaches to text retrieval. Technical report (1997)
19. Rossetto, L., Gasser, R., Heller, S., Amiri Parian, M., Schuldt, H.: Retrieval of structured and unstructured data with vitrivr. In: Proceedings of the ACM Workshop on Lifelog Search Challenge, LSC 2019, pp. 27–31. ACM, New York (2019). https://doi.org/10.1145/3326460.3329160
20. Zhou, B., Lapedriza, A., Khosla, A., Oliva, A., Torralba, A.: Places: a 10 million image database for scene recognition. IEEE Trans. Pattern Anal. Mach. Intell. **40**, 1452–1464 (2017)

Open Live Test for Question Retrieval

Final Report of the NTCIR-14 OpenLiveQ-2 Task

Makoto P. Kato[1](✉), Akiomi Nishida[2], Tomohiro Manabe[2], Sumio Fujita[2], and Takehiro Yamamoto[3]

[1] University of Tsukuba, Tsukuba, Japan
mpkato@slis.tsukuba.ac.jp
[2] Yahoo Japan Corporation, Tokyo, Japan
{anishida,tomanabe,sufujita}@yahoo-corp.jp
[3] University of Hyogo, Kobe, Japan
t.yamamoto@sis.u-hyogo.ac.jp

Abstract. This is the final report of the OpenLiveQ-2 task at NTCIR-14. This task aimed to provide an open live test environment of Yahoo Japan Corporation's community question-answering service (*Yahoo! Chiebukuro*) for question retrieval systems. The task was simply defined as follows: given a query and a set of questions with their answers, return a ranked list of questions. Submitted runs were evaluated both offline and online. In the online evaluation, we employed *pairwise preference multileaving*, a multileaving method that showed high efficiency over the other methods in a recent study. We describe the details of the task, data, and evaluation methods, and then report official results at NTCIR-14 OpenLiveQ-2. Furthermore, we demonstrate the effectiveness and efficiency of the proposed evaluation methodology.

Keywords: Online evaluation · Interleaving · Community question answering

1 Introduction

Community Question Answering (cQA) services are Internet services in which users can ask a question and obtain answers from other users. Users can obtain relevant information to their search intents not only by asking questions in cQA, but also by searching for questions that are similar to their intents. Finding answers to questions similar to a search intent is an important information seeking strategy especially when the search intent is very specific or complicated. While a lot of work has addressed the question retrieval problem [2,15,16], there are still several important problems to be tackled:

Ambiguous/underspecified queries. Most of the existing work mainly focused on specific queries. However, many queries used in cQA services are as short as Web search queries, and, accordingly, ambiguous/underspecified. Thus, question retrieval results also need diversification so that users with different intents can be satisfied.

© Springer Nature Switzerland AG 2019
M. P. Kato et al. (Eds.): NTCIR 2019, LNCS 11966, pp. 45–56, 2019.
https://doi.org/10.1007/978-3-030-36805-0_4

Diverse relevance criteria. The notion of relevance used in traditional evaluation frameworks is usually *topical relevance*, which can be measured by the degree of match between topics implied by a query and ones written in a document. Whereas, real question searchers have a wide range of relevance criteria such as freshness, concreteness, trustworthiness, and conciseness. Thus, traditional relevance assessment may not be able to measure real performance of question retrieval systems.

In order to address these problems, we have organized a task called *Open Live Test for Question Retrieval* (*OpenLiveQ*) since 2016, which provides an open live test environment of Yahoo! Chiebukuro[1] (a Japanese version of Yahoo! Answers) for question retrieval systems. Participants can submit ranked lists of questions for a particular set of queries, and receive evaluation results based on real user feedback. Involving real users in evaluation can solve problems mentioned above: we can consider the diversity of search intents and relevance criteria by utilizing real queries and feedback from users who are engaged in real search tasks.

The NTCIR-14 OpenLiveQ-2 task is the second round of OpenLiveQ [5]. The most of the settings in OpenLiveQ-2 are the same as those in the first round of OpenLiveQ (OpenLiveQ-1) [6]. We used the same task definition and the same query set for both training and testing, while we updated questions to be retrieved and clickthrough data in OpenLiveQ-2, and employed a new evaluation methodology for evaluating a large number of runs. In OpenLiveQ-1, only selected runs were evaluated in the online evaluation, since a prohibitively large amount of impressions were expected to statistically distinguish all the submitted runs. OpenLiveQ-2 tried to address this problem by proposing two-stage online evaluation [4]: the first stage identifies top-k runs and the second stage finds statistically significant differences only among top-k runs.

This final report includes the following topics:

1. The task definition of OpenLiveQ, which is used for both OpenLiveQ-1 and OpenLiveQ-2, and test collections developed for OpenLiveQ,
2. The evaluation results of the OpenLiveQ-2 task and approaches used by the task participants, and
3. The evaluation results of the proposed evaluation methodology.

The remainder of the paper is organized as follows. Section 2 describes the overview of the OpenLiveQ-2 task. Section 3 explains a new evaluation methodology applied to the OpenLiveQ-2 task. Section 4 briefly discusses the participants' approaches, and Sect. 5 presents the evaluation results. Section 6 concludes this paper with possible future directions.

2   Task

The task of OpenLiveQ-2 is simply defined as follows: given a query and a set of questions with their answers, return a ranked list of questions.

---

[1] http://chiebukuro.yahoo.co.jp/.

2.1 Phases

Our task consists of three phases:

1. **Offline Training Phase.** Participants were given *training data* including a list of queries, a set of questions for each query, and clickthrough data. They could develop and tune their question retrieval systems based on the training data.
2. **Offline Test Phase.** Participants were given only a list of queries and a set of questions for each query. They were required to submit a ranked list of questions for each query by a deadline. We evaluated submitted results by using evaluation metrics for ad-hoc retrieval with relevance judgment data that we developed in OpenLiveQ-1 [6]. Unlike OpenLiveQ-1, the results of the offline evaluation were only used for excluding poor ranking results that can drastically degrade the user satisfaction during the online evaluation. Meanwhile, we did not exclude any submitted runs in OpenLiveQ-2 since no run underperformed baseline runs to a large extent.
3. **Online Test Phase.** All the submitted runs were evaluated in a production environment of Yahoo Japan Corporation. A *multileaved comparison* method [8] was used in the online evaluation. As briefly mentioned in Sect. 1, OpenLiveQ-2 employed the two-stage online evaluation for evaluating a large number of runs efficiently.

2.2 Data

This section explains the data used in the OpenLiveQ-2 task.

Queries. The query set used in OpenLiveQ-2 is exactly the same as that in OpenLiveQ-1. The OpenLiveQ-1 queries were derived as follows. We first excluded time-sensitive and porn-related queries from a Yahoo! Chiebukuro search query log. The organizers then checked each of the queries and its questions, and filtered out a query if at least one of the organizers judged it had any of the ethic, discrimination, or privacy issues. Finally, we sampled 2,000 queries from the remaining queries, and used 1,000 queries for training and the rest for testing.

Some examples are "Bio Hazard"*, "Tibet"*, "Grape"*, "Prius"*, "twice", "Separate checks"*, and "gta5", where '*' indicates that a Japanese query was translated into English. It is worth noting that most of the queries consist of a single term: there are 1912 single-term queries, 68 two-term queries, and 20 three-term queries.

Questions. Questions were prepared in the same way as that in OpenLiveQ-1. We input each query to the current Yahoo! Chiebukuro search system as of Apr 10, 2018, recorded the top 1,000 questions, and used them as questions to be ranked. Information about all the questions as of Apr 10, 2018 was distributed to the OpenLiveQ participants. The total number of questions is 1,971,816. As was mentioned earlier, participants were required to submit a ranked list of those questions for each test query.

Clickthrough Data. Clickthrough data were collected in the same way as that in OpenLiveQ-1, and available for some of the questions. Based on the clickthrough data, one can estimate the click probability of the questions, and know what kinds of users click on a certain question. The clickthrough data were collected from Jan 10, 2018 to Apr 9, 2018. The number of query-question pairs in the clickthrough data is 436,890.

3 Evaluations

This section describes submissions from NTCIR-14 OpenLiveQ-2 participants, and then introduces the offline evaluation, in which runs were evaluated with relevance judgment data, and online evaluation, in which runs were evaluated with real users by means of multileaving.

3.1 Submissions

The NTCIR-14 OpenLiveQ-2 task attracted five research teams including an organizer team. The total number of submitted runs during the offline test phase was 65, of which 4 runs were duplicates of the other submissions. Thus, the total number of unique runs was 61.

3.2 Offline Evaluation

The offline evaluation was conducted in a similar way to traditional ad-hoc retrieval tasks, in which results are evaluated by relevance judgments and evaluation metrics such as nDCG (normalized discounted cumulative gain), ERR (expected reciprocal rank), and Q-measure. During the offline test period, participants could submit their results once per day through our Web site[2], and obtain evaluation results right after the submission.

While test questions used in OpenLiveQ-2 were not exactly the same as those in OpenLiveQ-1, we reused relevance judgment data in OpenLiveQ-2. The number of judged test questions was 43,205, *i.e.* 4.38% of all the test questions in OpenLiveQ-2. This fraction is comparable to that in OpenLiveQ-1, in which 4.54% of questions were judged. We used *condensed list* approach [10] to deal with incomplete relevance judgment data, *i.e.* we filtered out questions without relevance judgments from ranked lists of submitted runs.

The Q-measure score for each submitted run was displayed at our website. This is primarily because our recent study showed high correlation between the Q-measure scores and online evaluation results [4].

---

[2] http://www.openliveq.net/.

3.3   Online Evaluation

The NTCIR-13 OpenLiveQ-1 task attracted seven research teams and received 85 submissions in total. Even though the multileaved comparison can evaluate multiple rankings simultaneously, a large amount of search result impressions are required for a large number of rankers according to simulation-based experiments [8]. Thus, we opted to select a subset of submitted rankers by means of offline evaluation, and conducted multileaved comparison for only ten selected rankers—it turned out to be a problematic experimental design.

Lessons from NTCIR-13 OpenLiveQ task are summarized as follows [4]: (1) The offline evaluation results in terms of Q-measure [12] showed high correlation to the online evaluation results. However, there were some rankers for which the offline and online evaluation strongly disagreed. This implies a potential problem of our strategy: we might not evaluate rankers better than those selected for the online evaluation. This is a serious problem not only for an evaluation campaign but also for improvement of Web services. A straightforward solution to this problem is to evaluate all the rankers online. (2) A large number of users' clicks were necessary to find statistically significant differences for all the ranker pairs. As we cannot easily increase the number of search result impressions for multileaved comparison, a straightforward solution to this problem is to evaluate less rankers online.

These contradictory lessons motivated us to devise a new experimental design for large-scale multileaved comparison. Our proposed methodology in OpenLiveQ-2, two-stage online evaluation, is to evaluate all the rankers online for identifying top-k rankers, and intensively compare the top-k rankers so that they can get more chances to be statistically distinguished. We tested several top-k identification methods for multileaved comparison based on simulation experiments in our recent study [4]. The results demonstrated that even a simple method, Copeland counting algorithm, could achieve high recall in the top-k identification problem. Thus, OpenLiveQ-2 employed the two-stage online evaluation for evaluating all the submitted runs, with a recently proposed multileaving algorithm, pairwise preference multileaving (PPM) [8].

Given a set of rankings $\mathcal{R} = \{\mathbf{r}^{(1)}, \ldots, \mathbf{r}^{(n)}\}$ for a query, PPM selects one of $\Omega_r(\mathcal{R})$ with the uniform probability as a document at rank r in the combined ranking $\mathbf{m}$:

$$\Omega_r(\mathcal{R}) = \bigcup_{i=1}^{n} \mathbf{r}_{1:r}^{(i)} - \mathbf{m}_{1:r-1}, \tag{1}$$

where a ranking is defined as a sequence of documents, and $\mathbf{x}_{1:r}$ indicates a *set* of documents at rank $1, 2, \ldots, r$ in a ranking $\mathbf{x}$. In other words, PPM randomly select r-th document in the combined ranking from documents ranked at r-th rank or higher in rankings to be compared.

The purpose of multileaved comparison is to estimate a *preference matrix* P of which P_{ij} indicates how likely the i-th ranker is superior to the j-th ranker. PPM assigns a *credit* to each ranker based on observed clicks, and adds it to $\hat{P}_{ij}$

of an empirical preference matrix $\hat{P}$. After observing the feedback for a large number of search result impressions, we expected that P is approximated by $\hat{P}$.

Roughly speaking, PPM gives a positive credit to a ranker that agrees with pairwise preferences of documents inferred by the observed clicks, while a negative credit is given if a ranker disagree with them. PPM assumes that a clicked document is preferred to all of the unclicked documents above it, and the unclicked document at the next rank of the clicked document. A ranker is said to agree with "A is preferred to B" if the rank of A is higher than that of B in its ranking, and *vice versa*.

In our online evaluation, we evaluated only 61 unique runs out of 65 after excluding duplicate runs. The first stage of the two-stage online evaluation was carried out from Sep 28, 2018 to Nov 11, 2018. The total number of impressions used was 164,478 at the first stage. After identifying top-k runs based on the results of the first stage ($k = 30$ in OpenLiveQ-2), we evaluated only those top runs from Nov 23, 2018 to Jan 6, 2019. The total number of impressions used was 148,976 at the second stage.

4   Participants' Approaches

Four research teams submitted their runs at OpenLiveQ-2. This section briefly discusses their approaches.

4.1   AITOK [14]

AITOK tried to measure how catchy a question-answer pair is by using some statistics and query-question matching. The statistics include the number of views, number of answers, clickthrough rate, and update date. The matching degree was computed by TF-IDF, bigrams, and word embedding. These scores were combined together for finding catchy question-answer pairs, which were placed at the top of the ranking.

4.2   YJRS [7]

YJRS used BM25F features [9] as well as translation-based features [3] for learning to rank. BM25F was extended for handling numeric document fields and textual fields. The translation approach was proposed by one of the participants in OpenLiveQ-1, Erler, and translated best answer texts into model-generated question texts, from which basic features such as TF-IDF and BM25 were extracted. These features were combined by learning-to-rank algorithms. Note that this team submitted one of the best runs in OpenLiveQ-1 (Run ID: 92), and the baseline method based on learning-to-rank in OpenLIveQ-1 (Run ID: 91). Thus, they can be used to measure the progress of this task.

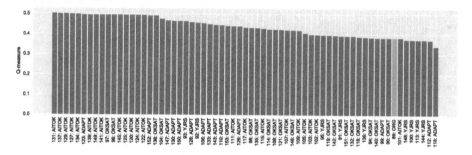

Fig. 1. Offline evaluation: Q-measure. (Color figure online)

4.3   OKSAT [13]

OKSAT utilized *white words* and *black words* for reranking questions. They defined nouns relevant to a query as white words, and nouns irrelevant to a query as black words. Their ranking algorithm ranks questions with more white words at higher ranks, and ones with more black words at lower ranks. They treated less frequent nouns in questions as black words, and as white words (1) frequent nouns in questions, (2) nouns suggested by Google suggestion, (3) manually chosen nouns, and (4) nouns in Wikipedia.

4.4   ADAPT [1]

ADAPT (or DCU) investigated the performances of different learning-to-rank models, feature selection and data normalization techniques. Coordinate ascent and MART were used for learning to rank. They categorized basic features such as TF-IDF and the number of answers into five classes, and tested the combination of these classes. They also explored several normalization techniques such as 0–1 normalization and standardization.

The four teams took fairly different approaches. AITOK used the word embedding technique and combined basic features manually. YJRS used some best features in the previous round and ranked questions by a learning-to-rank algorithm. OKSAT took a term-based approach that incorporated external knowledge sources such as Google and Wikipedia. ADAPT explored better combinations of important techniques for ranking tasks.

5   Evaluation Results

This section presents the offline and online evaluation results.

5.1   Offline Evaluation Results

Figures 1, 2, and 3 show results of the offline evaluation in terms of Q-measure, nDCG@10, and ERR@10. The baseline run (Run ID: 89) is indicated in red and

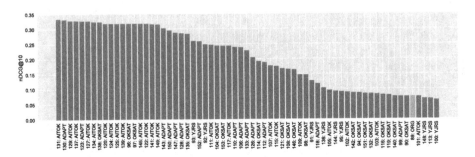

Fig. 2. Offline evaluation: nDCG@10. (Color figure online)

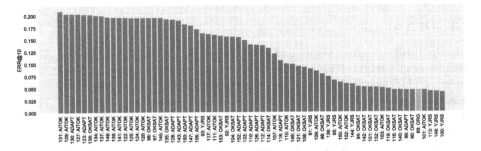

Fig. 3. Offline evaluation: ERR@10. (Color figure online)

was produced exactly the same ranked list as that used in the production. All the evaluation metrics show a similar trend as a whole, while the score of Q-measure is less sensitive to the runs than the others. AITOK and ADAPT performed well in the offline evaluation. It can be seen that the most runs outperformed the baseline run (Run ID: 89). It is also interesting to note that ID 92 (one of the best approach in OpenLiveQ-1) is not as effective as the online evaluation in OpenLiveQ-1.

5.2   Online Evaluation Results

Figures 4 and 5 show cumulated credits in the online evaluation at the first and second stage, respectively. Note that *the official result of NTCIR-14 OpenLiveQ-2 is that at the second stage, and online evaluation result at the first stage is only considered as unofficial due to lack of statistical power.* Looking at the top performers in Fig. 5, YJRS performed well and AITOK as well as ADAPT did not as they did in the offline evaluation. This is discussed further in the next subsection.

5.3   Evaluation of Evaluation Methodology

Since we employed a new evaluation methodology in this round, we evaluated our evaluation methodology from several aspects.

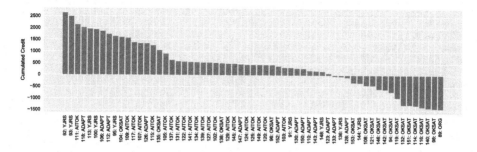

Fig. 4. Online evaluation at the first stage.

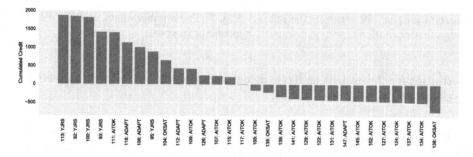

Fig. 5. Online evaluation at the second stage.

Is There a High Correlation Between the Offline and Online Evaluations? To answer this question, we first tried to quantify the difference of run pairs by using the whole data in the two stages. The credits shown in Figs. 4 and 5 cannot be simply used together, as their scales can be different due to different ranking sets in the first and second stages. Thus, we used the difference of the number of *wins* and *loses* for pairs of runs for quantifying the difference, where *win* means a run received a higher credit in an impression than the other, and *lose* means a run received a lower credit in an impression than the other (ties were ignored). The difference in the offline evaluation was simply defined as the Q-measure difference.

Figure 6 shows the run pair difference in the offline and online evaluations. The evaluation results in those phases show a negative correlation: Pearson's $r = -0.488$. This result supported one of the lessons from NTCIR-13 OpenLiveQ task: there are some rankers for which the offline and online evaluations strongly disagreed. Possible explanations for the strong disagreement between the offline and online evaluations are (1) the relevance judgements were not so similar to real users' clicks, (2) the relevance of some questions were not judged in OpenLiveQ-2, and (3) systems were heavily tuned to the relevance judgements, *i.e.* over-fitting.

The difficulty of predicting online evaluation results can be a strong motivation to conduct online evaluations, and may require additional studies on the difference of the two types of evaluations. Meanwhile, it is also a strong

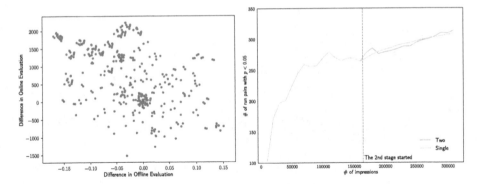

Fig. 6. Run pair difference in the offline and online evaluations.

Fig. 7. The number of statistically distinguished run pairs.

motivation for the evaluation methodology we employed, *i.e.* all the runs should be evaluated online.

Is the Two-Stage Strategy Better Than the Single-Stage Strategy?
In contrast to the two-stage strategy, single-stage strategy means evaluating all the runs for the entire period. Since the single-stage strategy was not used in OpenLiveQ-2, we simulated an extension of the first stage by repeatedly sampling impressions during the first stage. As a result, we obtained two sets of impressions: one comprising impressions from the first and second stages, and one comprising impressions from the first stage and samples from the first stage. The number of impressions is exactly the same, *i.e.* $164,478+148,976 = 313,454$. Since the sampling is probabilistic, we repeated the simulation 100 times and report the average of 100 simulations below.

Since the two-stage strategy was employed for a higher discriminative power, we counted the number of statistically distinguished run pairs by using the two types of impression sets. For the same reason as the previous analysis, we used wins and loses of run pairs for significant tests. The Wilcoxon signed-rank test was used with significance level $\alpha = 0.05$ and Holm correction to deal with the multiple comparison problem. Figure 7 shows the number of statistically distinguished run pairs as a function of the number of impressions. When the number of impressions is smaller than 164,478, the two strategies produced exactly the same result since the impressions are identical. With more impressions than 164,478, the discriminative power, or the number of statistically distinguished run pairs, is more stable for the single-stage strategy as it is the average of 100 simulations. When all the impressions were used, the discriminative power for the two-stage strategy is slightly higher than that of the single-stage strategy. However, it is not very conclusive that the two-stage strategy has a higher discriminative power than the single-stage strategy.

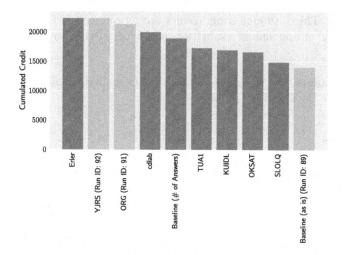

Fig. 8. Online evaluation result at OpenLiveQ-1.

Is the Interleaving Evaluation Reproducible? Reproducibility is an important challenge for online evaluation. We investigated whether our interleaving evaluation in OpenLiveQ-2 reproduced the run differences reported in OpenLiveQ-1, by following the idea of NTCIR-14 CENTRE [11].

Figure 8 shows online evaluation results in OpenLiveQ-1, which is derived from Fig. 3 of the NTCIR-13 OpenLiveQ-1 overview paper [6]. As mentioned earlier, YJRS submitted runs that were produced by exactly the same approaches as those in OpenLiveQ-1. More concretely, "YJRS" in Fig. 8 corresponds to 92 in OpenLiveQ-2, "ORG" corresponds to 91, and "Baseline (as is)" corresponds to 89. Thus, when A outperformed B in OpenLiveQ-1, their difference is reproduced if A outperforms B in OpenLiveQ-2. Comparing Figs. 4 and 8, we can observe that all the differences were reproduced. Therefore, though the number of samples is small, OpenLIveQ-2 could successfully reproduced some results in OpenLiveQ-1.

6   Conclusion

This paper reported the task design, evaluation methodology, evaluation results, and in-depth analysis of the evaluation methodology in the NTCIR-14 OpenLiveQ-2 task. We found that (1) there is a strong disagreement between offline and online evaluations, (2) the most OpenLiveQ-2 runs outperformed the baseline ranking method used in the production system, (3) a small improvement of the discriminative power was achieved by the two-stage strategy, and (4) OpenLIveQ-2 could successfully reproduced some results in OpenLiveQ-1.

There are still many challenges in the online evaluation. First, it is worth investigating the difference between offline and online evaluations. Second, a large-scale online evaluation is still challenging and may need more studies with many rankers. Third, though some results are reproduced in OpenLiveQ-2, the reproducibility of the online evaluation could be further investigated in the future.

Acknowledgments. We would like to thank the OpenLiveQ-2 participants for their contributions to the OpenLiveQ-2 task.

References

1. Arora, P., Jones, G.: DCU at the NTCIR-14 OpenLiveQ-2 task. In: NTCIR-14 Conference (2019)
2. Cao, X., Cong, G., Cui, B., Jensen, C.S.: A generalized framework of exploring category information for question retrieval in community question answer archives. In: WWW, pp. 201–210 (2010)
3. Chen, M., Li, L., Sun, Y., Zhang, J.: Erler at the NTCIR-13 OpenLiveQ task. In: NTCIR-13 Conference (2017)
4. Kato, M.P., Manabe, T., Fujita, S., Nishida, A., Yamamoto, T.: Challenges of multileaved comparison in practice: lessons from NTCIR-13 OpenLiveQ task. In: CIKM, pp. 1515–1518 (2018)
5. Kato, M.P., Nishida, A., Manabe, T., Fujita, S., Yamamoto, T.: Overview of the NTCIR-14 OpenLiveQ-2 task. In: NTCIR-14 Conference (2019)
6. Kato, M.P., Yamamoto, T., Manabe, T., Nishida, A., Fujita, S.: Overview of the NTCIR-13 OpenLiveQ task. In: NTCIR-13 Conference (2017)
7. Manabe, T., Fujita, S., Nishida, A.: YJRS at the NTCIR-14 OpenLiveQ-2 task. In: NTCIR-14 Conference (2019)
8. Oosterhuis, H., de Rijke, M.: Sensitive and scalable online evaluation with theoretical guarantees. In: CIKM, pp. 77–86 (2017)
9. Robertson, S., Zaragoza, H., Taylor, M.: Simple BM25 extension to multiple weighted fields. In: CIKM, pp. 42–49 (2004)
10. Sakai, T.: Alternatives to Bpref. In: SIGIR, pp. 71–78 (2007)
11. Sakai, T., Ferro, N., Soboroff, I., Zeng, Z., Xiao, P., Maistro, M.: Overview of the NTCIR-14 centre task. In: NTCIR-14 Conference (2019)
12. Sakai, T., Song, R.: Evaluating diversified search results using per-intent graded relevance. In: SIGIR, pp. 1043–1052 (2011)
13. Sato, T., Nagase, Y., Uraji, M.: OKSAT at NTCIR-14 OpenLiveQ-2 task -reorder questions by using white and black words. In: NTCIR-14 Conference (2019)
14. Tanioka, H.: AITOK at the NTCIR-14 OpenLiveQ-2 task. In: NTCIR-14 Conference (2019)
15. Wang, K., Ming, Z., Chua, T.S.: A syntactic tree matching approach to finding similar questions in community-based QA services. In: SIGIR, pp. 187–194 (2009)
16. Zhou, G., Liu, Y., Liu, F., Zeng, D., Zhao, J.: Improving question retrieval in community question answering using world knowledge. In: IJCAI, pp. 2239–2245 (2013)

Online Evaluations of Features and Ranking Models for Question Retrieval

Tomohiro Manabe[✉], Sumio Fujita, and Akiomi Nishida

Yahoo Japan Corporation, Chiyoda, Tokyo 102-8282, Japan
{tomanabe,sufujita,anishida}@yahoo-corp.jp

Abstract. We report our work on the NTCIR-14 OpenLiveQ-2 task. From the given data set for question retrieval on a community QA service, we extracted several BM25F-like features and translation-based features in addition to basic features such as TF, TFIDF, and BM25 and then constructed multiple ranking models with the feature sets. In the first stage of online evaluation, our linear models with the BM25F-like and translation-based features obtained the highest amount of credit among 61 methods including other teams' methods and a snapshot of the current ranking in service. In the second stage, our neural ranking models with basic features consistently obtained a major amount of credit among 30 methods in a statistically significant high number of page views. These online evaluation results demonstrate that neural ranking is one of the most promising approaches to improve the service.

Keywords: Multileaved comparison · Neural ranking model · Question retrieval · Linear combination model · Ensemble tree model

1 Introduction

We report our work on the NTCIR-14 OpenLiveQ-2 task [7], where runs, *i.e.* search result rankings of our question retrieval systems, have been evaluated, together with other participants', using a multileaving strategy. All runs were evaluated online at Yahoo! *Chiebukuro*[1], a community QA service in Japanese.

As our work in the previous round of the task, NTCIR-13 OpenLiveQ-1 [10], we trained a linear combination of 77 basic features with the Coordinate Ascent (CA) method [11]. Then we tried to add some features, namely, BM25F-like features of the existing document fields. In OpenLiveQ-1, this approach achieved the best nDCG@10 score among our runs, thus selected to be evaluated in online evaluation. However, it is not clear that other runs, irrespective of performing better in offline evaluation or not, might outperform this approach in online evaluation. Thanks to the new multileaved comparison approach, we take the opportunity of the current round where most submitted runs are evaluated online in order to carry out more thorough investigations. Thus we try to investigate our

---

[1] https://chiebukuro.yahoo.co.jp/.

© Springer Nature Switzerland AG 2019
M. P. Kato et al. (Eds.): NTCIR 2019, LNCS 11966, pp. 57–69, 2019.
https://doi.org/10.1007/978-3-030-36805-0_5

Table 1. List of our runs and their offline evaluation results.

ID	Description	Q-measure	nDCG@10	ERR@10
91	CA+Basic	0.39124	0.12667	0.08849
92	CA+BM25F	0.45609	0.24802	0.15548
93	CA+Trans	0.46387	0.25926	0.16244
95	CA+Basic (retry)	0.39559	0.08976	0.06469
100	ListNet+Basic	0.37340	0.06296	0.03971
113	ListNet+Basic with 5-fold cross validation	0.37240	0.06660	0.04225
136	CA+All	0.38514	0.10256	0.07252
144	LightGBM+Basic	0.37228	0.09128	0.05689
148	LightGBM+Basic with rough parameter tuning	0.37429	0.06710	0.04141

Table 2. Matrix of p-values for statistical test for difference between mean Q-measure scores of two of our runs according to Student's paired t-test.

	91	92	93	95	100	113	136	144	148
91	–	.0015	.0004	.8301	.3680	.3419	.7663	.3452	.3978
92	.0015	–	.6907	.0021	.0000	.0000	.0004	.0000	.0000
93	.0004	.6907	–	.0006	.0000	.0000	.0001	.0000	.0000
95	.8301	.0021	.0006	–	.2466	.2262	.5990	.2303	.2719
100	.3680	.0000	.0000	.2466	–	.9570	.5446	.9526	.9622
113	.3419	.0000	.0000	.2262	.9570	–	.5110	.9950	.9199
136	.7663	.0004	.0001	.5990	.5446	.5110	–	.5129	.5802
144	.3452	.0000	.0000	.2303	.9526	.9950	.5129	–	.9161
148	.3978	.0000	.0000	.2719	.9622	.9199	.5802	.9161	–

research questions (RQ) in suspense throughout the previous round: (RQ1) are there any feature sets which may further improve the online effectiveness? (RQ2) Are there any ranking models that outperform CA in the online evaluation?

For RQ1, we appended features of *best answer texts in question language*, which are best answer texts translated into pseudo question texts with a translation model and examined four feature sets in total, whereas the ranking model is fixed to CA. For RQ2, we also tried to replace the linear combination with other widely used models, namely, (1) a neural ranking model generated with ListNet [2] and (2) Gradient Boosting Decision Trees (GBDT) generated with LightGBM [8], whereas the feature set is fixed to the basic set.

The remainder of the paper is organized as follows: Sect. 2 explains each examined runs. Sections 3 and 4 describe the results of offline and online evaluations respectively. Finally Sect. 5 concludes the paper.

2   Examined Feature Sets and Ranking Models

In this section, we explain our submitted runs describing the adopted feature sets and ranking models. Table 1 lists all our runs and their offline evaluation results in temporal order of submission. Table 2 is a matrix of p-values for the statistical test for the difference between the mean scores of two of our runs according to Student's paired t-test. Each pair consists of Q-measure scores, the official primary evaluation measure, against one test query.

2.1   Linear Combination Model of Basic Features (CA+Basic)

The task organizers provided us with the data set and a tool[2] for basic feature extraction from the data set. The tool's README file includes a short instruction for generating a simple linear combination of the features by using the RankLib implementation[3] of CA [11]. First we followed the instruction for generating a linear combination of 77 basic features (Run ID: 91).

Among the 77 features, 68 are composed of 17 feature types (TF, IDF, ICF[4], TFIDF, TFICF, BM25, language models with three smoothing methods, document length, and their logarithmic and/or normalized variations), most of which are common to the well-known LETOR data set [14], extracted from each of four textual fields (question title, question snippets, question text, and related best answer text). The other nine features are answer count, page view count, their logarithmic variations, average rank at the current rankings in service, timestamp, and three 0/1 flags (open to answer, open to vote for best answer, and solved). We calculated feature values with our original Solr[5] plug-in instead of the official tool because of its expandability. This is the only difference between the official instruction and the generation process of this model.

We used the RankLib implementation of CA for learning a linear combination model from the training data composed of 1,000 queries and 986,125 questions in total. The CA method optimizes each parameter one-by-one. To optimize a parameter, it examines some smaller and larger points from the current value and greedily adopts a new value that improves the objective function. After optimizing the last parameter, it shuffles and iterates over all the parameters again. The optimization finishes if modification of no parameter can improve the value of the objective function. As its objective function, we used the default ERR@10 [3]. As relevance judgment labels between queries and questions, we naively normalized given click-through rates. For the normalization, we divided the rates by the max for the query, multiplied them by four (max relevance grade), and then truncated them to integers. We call this the *CA+Basic* method.

Because CA is a probabilistic process and the nDCG@10 score of Run 91 was worse than the previous task [10], we attempted to generate another CA+Basic run (Run ID: 95); however, that did not improve nDCG@10.

---

[2] https://github.com/mpkato/openliveq.
[3] https://sourceforge.net/p/lemur/wiki/RankLib/.
[4] |documents in collection|/|keyword occurrences in collection|.
[5] https://lucene.apache.org/solr/.

2.2   Extended BM25F Features (CA+BM25F)

In the previous NTCIR-13 OpenLiveQ task [6], we proposed an extension of the CA+Basic method [10]. Because of its simplicity and its top-performance in the previous task, we also generated and submitted a run for this task with the method (Run ID: 92). Its differences from CA+Basic are as follows:

- In addition to the 77 basic features, we use three BM25F [15] features extended for handling numeric document fields as well as textual fields. The three features are based on three different field weighting strategies.
- We use nDCG@10 as the objective function of CA.
- We perform 5-fold cross validation on the training data.

In the last task, we had assigned negative BM25F weights to some fields, while in this task, we used zero instead because a harmful effect of negative weights was found. We call this the *CA+BM25F* method.

2.3   Translation Features (CA+Trans)

In addition to CA+BM25F, a method based on a translation model outperformed the other runs in the last task [4]. Its key idea is different language models behind queries, questions, and answers. On the basis of the following two observations, it translates answers into questions: (1) Queries must be more similar to questions than answers, and (2) there is enough data for constructing translation models between questions and answers. We also adopted this idea for this task; namely, we translated best answer texts into model-generated question texts and extracted 17 features of textual fields (discussed in Sect. 2.1) from the resulting text.

The translation was based on a translation model from answer texts to question texts easily constructed with the GIZA++ toolkit [12] and the publicly available Yahoo! *Chiebukuro* corpus[6]. The translation model is a set of correspondences between one answer term and multiple question terms with their translation probabilities, e.g., fruit → apple (50%), banana (30%), and orange (20%). To translate an answer text into a single (not probabilistic) question text, we naively iterated over the answer text cumulating the probabilities for each question term as the term's score, sorted the terms by score, and then extracted the top-l terms where l is the number of term occurrences in the answer text. We generated a run by linearly combining the 94 features in total (77 basic features and 17 features of model-generated text) with CA and submitted it (Run ID: 93). We call this the *CA+Trans* method.

2.4   Combination of Extended BM25F and Translation (CA+All)

Intuitively, we can independently apply the modifications explained in Sects. 2.2 and 2.3. We also generated a run that incorporates both modifications (Run ID: 136). We call this the *CA+All* method.

---

[6] https://www.nii.ac.jp/dsc/idr/yahoo/chiebkr2/Y_chiebukuro.html.

2.5   Neural Ranking Model (ListNet+Basic)

Neural ranking is a state-of-the-art approach to document scoring for retrieval. Among a wide variety of neural ranking methods, we used ListNet [2] for combining the basic 77 features (Run ID: 100). According to the paper, the key concept of ListNet is its listwise loss function which has advantage against pairwise loss functions. As the ranking model, we used a simple 3-layered fully connected feed-forward neural network whose size of hidden layer is 200. We normalized the feature values into [0, 1] with simple min-max normalization. We used Chainer[7] along with its Adam optimizer implementation [9] (initial learning rate: 0.0007; learning rate decay factor: 0.995) as our neural ranking framework. We set the batch size to 512, the number of iterations to 1,000, and the weight decay factor to 0.0005. Further tuning of hyper-parameters with grid search did not improve the offline evaluation results significantly. We call this the *ListNet+Basic* method.

We also applied 5-fold cross validation to this approach (Run ID: 113).

Although we could input feature sets other than the basic one to the neural ranking models, we chose not to do so due to the time limitation of this task.

2.6   Ensemble Tree Model (LightGBM+Basic)

The GBDT is another state-of-the-art approach to document scoring for retrieval. We used the LightGBM implementation [8] of GBDT and the nDCG-based LambdaRank objective function [1] for generating a run by combining the 77 basic features (Run ID: 144). The model of GBDT is ensemble trees, i.e., linear combination of regression trees. The GBDT generates trees one-by-one for minimizing errors on already generated trees with gradient descent. We learned an ensemble of 100 trees including 15 leaves, each with a learning rate of 0.1. The LightGBM implementation supports other techniques, including feature binning, bagging, pruning, and so on. We set the max number of bins to 255, the bagging fraction to 0.9, the minimum number of data points per leaf to 50, and the minimum summation of Hessians per leaf to 5.0. We call this the *LightGBM+Basic* method.

We also tried to improve its effectiveness by a simple grid search over its hyper-parameter space; however, that did not improve the score (Run ID: 148).

As with the neural ranking models, we could input feature sets other than the basic one to LightGBM, but we chose not to do so.

3   Offline Evaluation Results

We already listed all our runs' scores on three evaluation measures in Table 1 and the statistical significance of their differences on Q-measure in Table 2.

Overall, the offline evaluation results were not stable. The nDCG and ERR scores of runs 91 and 95 are, for example, very different, even though these runs were generated with exactly the same (but probabilistic) CA+Basic method.

---

[7] https://chainer.org/.

In contrast, the Q-measure produced relatively stable scores throughout the offline evaluation results. This fact supports the validity of selecting Q-measure as the official primary evaluation measure of this offline evaluation. According to this measure, the CA+BM25F (Run ID: 92) and CA+Trans (Run ID: 93) methods were significantly effective among our runs. This demonstrates the utility of the extended BM25F and translation-based features. There was no significant difference among the other runs. From the viewpoint of statistical testing, runs 92 and 93 also significantly outperformed our other runs on Q-measure ($p < 0.005$). Between the two runs, no statistically significant difference was found. Similarly, among the other runs, no statistically significant difference was found.

4   Online Evaluation Results

In this section, we discuss the online evaluation results with a focus on our runs.

The key idea of pairwise multileaved comparison is that, when a user submits a test query to the service, the system looks up corresponding rankings from runs under the comparison. After that, it interleaves the rankings into a mixed ranking and then returns it to the user. If the user clicks an item on the mixed ranking, it means that the user prefers the clicked item to any unclicked items above that. Therefore, for each of the pairwise preferences, some credit is given to the runs that rank the clicked item higher than the unclicked item.

In this task, the online evaluation period consists of two stages of multileaved comparisons [5] with Pairwise Preference Multileaving [13] on Yahoo! *Chiebukuro*. We treat the two as independent experiments, i.e., we do not sum up the credits, as it is difficult to assign reasonable weights to each of the stages.

4.1   First Stage

This stage of the online evaluation was conducted from Sept. 28 to Nov. 11, 2018. In this stage, 164,478 page views were made on mixed rankings.

Figure 1 shows the credit of our runs (marked with *) and the other teams' best (in credit) runs in this evaluation stage. Note that Run 89 is a snapshot of the ranking in service submitted by the task organizers as a baseline. Table 3 is a matrix of p-values for the statistical test for the difference between the credits of two runs according to Student's paired t-test. Each pair consists of the credit of two runs against one test query. In total, 61 runs were compared, but we have omitted the other 48 due to space limitations.

Table 4 counts page views in this evaluation stage when the run arranged in the row obtained a larger credit than the run arranged in the column. Underlines mean that the run arranged in the row obtained a larger credit than the run arranged in the column in more than 50% of counts. Table 5 lists p-values for the statistical test for the difference between winning percentages of two runs according to a binomial test.

As shown in Fig. 1, our CA+BM25F (Run ID: 92) and CA+Trans (Run ID: 93) runs obtained the highest credit in total among all the runs. This result is

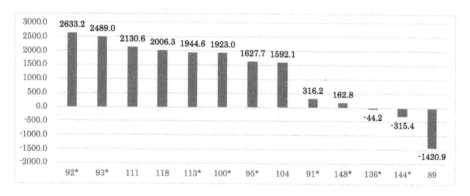

Fig. 1. First-stage online evaluation results of our runs (marked with *) and other teams' best runs in descending order of credit. X-axis is run ID and y-axis is credit.

Table 3. Matrix of p-values for statistical test for difference between credit amount of two runs in Fig. 1 according to Student's paired t-test.

	92*	93*	111	118	113*	100*	95*	104	91*	148*	136*	144*	89
92*	–	.6906	.2320	.0927	.0681	.0623	.0046	.0058	.0000	.0000	.0000	.0000	.0000
93*	.6906	–	.3887	.1891	.1435	.1319	.0138	.0159	.0000	.0000	.0000	.0000	.0000
111	.2320	.3887	–	.7700	.6647	.6310	.2195	.2093	.0000	.0000	.0000	.0000	.0000
118	.0927	.1891	.7700	–	.8717	.8290	.2934	.2784	.0000	.0000	.0000	.0000	.0000
113*	.0681	.1435	.6647	.8717	–	.9560	.3855	.3619	.0000	.0000	.0000	.0000	.0000
100*	.0623	.1319	.6310	.8290	.9560	–	.4231	.3961	.0000	.0000	.0000	.0000	.0000
95*	.0046	.0138	.2195	.2934	.3855	.4231	–	.9222	.0003	.0000	.0000	.0000	.0000
104	.0058	.0159	.2093	.2784	.3619	.3961	.9222	–	.0009	.0001	.0000	.0000	.0000
91*	.0000	.0000	.0000	.0000	.0000	.0000	.0003	.0009	–	.6793	.3454	.0891	.0000
148*	.0000	.0000	.0000	.0000	.0000	.0000	.0000	.0001	.6793	–	.5788	.1866	.0000
136*	.0000	.0000	.0000	.0000	.0000	.0000	.0000	.0000	.3454	.5788	–	.4672	.0004
144*	.0000	.0000	.0000	.0000	.0000	.0000	.0000	.0000	.0891	.1866	.4672	–	.0035
89	.0000	.0000	.0000	.0000	.0000	.0000	.0000	.0000	.0000	.0000	.0004	.0035	–

consistent with the offline evaluation results. However, from the viewpoint of statistical testing (see Fig. 3), these top two run's credits were not significantly larger than the other top-tiers, runs 111, 118, 113, and 100. Compared to runs 95 and below, these top two run's credits were significantly higher ($p < 0.05$).

According to Tables 4 and 5, the winning percentages of these top two runs were consistently over 50% against the other runs, and their superiorities are statistically significant ($p < 0.05$). Between these top two runs, the CA+BM25F run obtained a slightly higher amount of credit in total while the CA+Trans run obtained more credits than the CA+BM25F one in more page views. However, the differences were not statistically significant.

Table 4. First-stage win-lose page view counts of our runs (marked with *) and other teams' best runs. Each element is count of page views where run arranged in row obtained larger credit than run arranged in column. Underlines mean that winning percentage of run arranged in row is higher than 50%.

	92*	93*	111	118	113*	100*	95*	104	91*	148*	136*	144*	89
92*	–	2873	6092	4612	5792	5771	4178	6166	6966	6973	7306	7038	7509
93*	3021	–	5856	5084	5706	5693	4226	5912	6809	6849	7176	6933	7413
111	5716	5307	–	4440	5750	5733	5532	4757	6857	6902	7246	6934	7443
118	4385	4716	4676	–	5753	5695	5162	5570	6956	6980	7266	7072	7425
113*	5523	5319	5881	5744	–	489	4726	6303	4554	4432	5102	5044	6116
100*	5503	5322	5851	5701	492	–	4730	6311	4631	4494	5173	5068	6128
95*	3926	3813	5650	5087	4701	4714	–	5919	5827	6027	6492	6201	6839
104	5856	5520	4827	5371	6312	6289	5959	–	7126	7136	7436	7147	7435
91*	5790	5552	6066	6034	3406	3482	4876	6220	–	2820	2291	3101	3975
148*	5838	5647	6175	6075	3319	3382	5071	6261	2935	–	2891	1418	4000
136*	5898	5663	6167	6100	3716	3778	5258	6261	2001	2461	–	2578	2793
144*	5809	5566	6109	6038	3876	3912	5118	6143	3076	1315	2885	–	3875
89	5774	5585	6066	5938	4431	4475	5265	5964	3301	3222	2430	3235	–

Table 5. Matrix of p-values for the statistical test for difference between winning percentage of two runs in Table 4 according to binomial test.

	92*	93*	111	118	113*	100*	95*	104	91*	148*	136*	144*	89
92*	–	.0555	.0006	.0172	.0118	.0119	.0053	.0048	.0000	.0000	.0000	.0000	.0000
93*	.0555	–	.0000	.0002	.0002	.0004	.0000	.0003	.0000	.0000	.0000	.0000	.0000
111	.0006	.0000	–	.0138	.2280	.2770	.2685	.4809	.0000	.0000	.0000	.0000	.0000
118	.0172	.0002	.0138	–	.9405	.9626	.4648	.0584	.0000	.0000	.0000	.0000	.0000
113*	.0118	.0002	.2280	.9405	–	.9491	.8048	.9432	.0000	.0000	.0000	.0000	.0000
100*	.0119	.0004	.2770	.9626	.9491	–	.8773	.8516	.0000	.0000	.0000	.0000	.0000
95*	.0053	.0000	.2685	.4648	.8048	.8773	–	.7205	.0000	.0000	.0000	.0000	.0000
104	.0048	.0003	.4809	.0584	.9432	.8516	.7205	–	.0000	.0000	.0000	.0000	.0000
91*	.0000	.0000	.0000	.0000	.0000	.0000	.0000	.0000	–	.1329	.0000	.7601	.0000
148*	.0000	.0000	.0000	.0000	.0000	.0000	.0000	.0000	.1329	–	.0000	.0510	.0000
136*	.0000	.0000	.0000	.0000	.0000	.0000	.0000	.0000	.0000	.0000	–	.0000	.0000
144*	.0000	.0000	.0000	.0000	.0000	.0000	.0000	.0000	.7601	.0510	.0000	–	.0000
89	.0000	.0000	.0000	.0000	.0000	.0000	.0000	.0000	.0000	.0000	.0000	.0000	–

As shown in Fig. 1, among our runs, the ListNet+Basic with 5-fold cross validation (Run ID: 113) and simple ListNet+Basic (Run ID: 100) runs followed the top two runs. Between these second-tier runs, there was no significant difference. This indicates that optimization of neural ranking models by cross validation was not necessary. It seems that our learning process generates models with stable performance, or that run 100 was as well-trained as run 113 by chance.

Our CA+Basic runs (Run IDs: 95 and 91) occupied the fifth and sixth positions among our runs. Although they are based on the same method, their credit amounts are statistically significantly different ($p < 0.0005$, see Fig. 1 and Table 3). We assume this is due to the unstableness of CA. In this context, the unstableness means the accuracy of linear combination models generated with CA varies widely. This hypothesis also explains the wide range of credit and win-lose counts of our linear combination models (Runs 92, 93, 95, 91, and 136).

There was a large gap of total credit between runs 104 and 91 (see Fig. 1). All the runs above the gap obtained statistically significant larger credits than the runs below ($p < 0.001$, see Table 3) and outperformed the runs below in statistically significant large numbers of page views ($p < 0.00005$, see Table 5).

As explained above, our GBDT+Basic runs with and without parameter tuning (Run IDs: 148 and 144) are below the large gap of total credit. In other words, our GBDT runs did not perform as well as other runs in this comparison. We conclude this is because of the small amount of training data.

The snapshot of the current ranking in service (Run ID: 89) obtained only the smallest amount of credit among all 61 runs in this evaluation stage (see Fig. 1). Moreover, the difference from all the other runs in Table 3 is statistically significant ($p < 0.005$).

4.2 Second Stage

This stage of online evaluation was conducted from Nov. 23, 2018 to Jan. 6, 2019. During this time, 148,976 page views were made on mixed rankings.

The same as the first stage, we present the total credits (Fig. 2), a matrix of p-values according to paired t-test (Table 6), win-lose page view counts (Table 7), and a matrix of p-values according to binomial tests (Table 8).

In this stage, 30 runs were compared. These runs were selected mainly based on the total credit obtained in the first stage. They are the runs above the large gap of total credits in the first stage.

As indicated in the figure and tables, among our runs, ListNet+Basic with and without 5-fold cross validation (Run IDs: 113 and 100), CA+BM25F (Run ID: 92), CA+Trans (Run ID: 93), and CA+Basic (Run ID: 95) passed the first stage. As a result, each run had a greater chance of obtaining credit in this stage than in the first stage. In other words, we obtained about 5,000 page views per run in this stage of comparison, in contrast to the 2,700 per run in the previous stage. For this reason, we consider the results of this stage to be more reliable than those of the first stage.

Comparing the evaluation results of the second stage with those of the first stage, our ListNet+Basic with and without 5-fold cross validation (Run IDs: 113 and 100) occupied better positions (compare Figs. 1 and 2). We assume this is due to the more accurate comparison. Between these two runs, there was no significant or statistically significant difference in terms of credits or win-lose

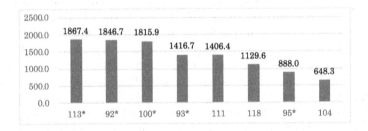

Fig. 2. Second-stage online evaluation results of our runs (marked with *) and other teams' best runs in descending order of credit. X-axis is run IDs and y-axis is credit.

Table 6. Matrix of p-values of statistical test for difference between second-stage credits of two runs in Fig. 2 according to Student's paired t-test.

	113*	92*	100*	93*	111	118	95*	104
113*	–	.9619	.9068	.3115	.3867	.0929	.0334	.0148
92*	.9619	–	.9429	.3245	.4018	.0954	.0338	.0149
100*	.9068	.9429	–	.3677	.4404	.1161	.0428	.0191
93*	.3115	.3245	.3677	–	.9848	.5157	.2533	.1261
111	.3867	.4018	.4404	.9848	–	.6010	.3434	.1919
118	.0929	.0954	.1161	.5157	.6010	–	.5967	.3324
95*	.0334	.0338	.0428	.2533	.3434	.5967	–	.6418
104	.0148	.0149	.0191	.1261	.1919	.3324	.6418	–

Table 7. Second-stage win-lose page view counts of our runs (marked with *) and other teams' best runs. Each element is count of page views where run arranged in row obtained higher credit than run arranged in column. Underlines mean that winning percentage of run arranged in row is higher than 50%.

	113*	92*	100*	93*	111	118	95*	104
113*	–	6433	440	6231	6662	6700	5475	7114
92*	6131	–	6093	3449	6864	5502	4703	6963
100*	451	6426	–	6235	6674	6694	5478	7127
93*	5919	3612	5923	–	6506	5917	4732	6650
111	6036	6630	6026	6194	–	5386	6167	5313
118	6100	5242	6061	5547	5381	–	5867	6247
95*	5012	4658	5023	4561	6362	6006	–	6693
104	6656	6705	6653	6404	5277	6223	6550	–

Table 8. Matrix of p-values of statistical test for difference between winning percentages of two runs in Table 7 according to binomial test.

	113*	92*	100*	93*	111	118	95*	104
113*	–	.0072	.7376	.0048	.0000	.0000	.0000	.0001
92*	.0072	–	.0030	.0539	.0449	.0125	.6493	.0279
100*	.7376	.0030	–	.0048	.0000	.0000	.0000	.0001
93*	.0048	.0539	.0048	–	.0058	.0006	.0778	.0320
111	.0000	.0449	.0000	.0058	–	.9693	.0831	.7338
118	.0000	.0125	.0000	.0006	.9693	–	.2053	.8368
95*	.0000	.6493	.0000	.0778	.0831	.2053	–	.2172
104	.0001	.0279	.0001	.0320	.7338	.8368	.2172	–

page view counts. This fact again indicates the stable performance of our learning process of neural ranking models. From the viewpoint of statistical testing, credit amounts did not have enough statistical power to distinguish these two runs from the others (see Table 6); however, they outperformed the others in statistically significant large numbers of page views ($p < 0.01$, see Table 8).

Interestingly, our CA+BM25F run (Run ID: 92) obtained an amount of credits comparable with our ListNet+Basic runs. However, as mentioned above, it obtained a smaller amount of credit than our ListNet+Basic runs in statistically significant large numbers of page views ($p < 0.01$, see Table 8). This fact indicates that our CA+BM25F run obtained quite a large amount of credit in only a few page views, whereas our ListNet+Basic runs consistently obtained credit.

Other tendencies were almost the same as in the first stage. More precisely, the order among the runs, except for our ListNet runs, was the same.

5   Concluding Remarks

We examined four feature sets and three ranking models in the NTCIR-14 OpenLiveQ-2 task. The feature sets include (1) basic features such as TF, TFIDF, and BM25 (Basic), (2) basic features with extended BM25F features (BM25F), (3) basic features with translation-based features extracted from pseudo questions generated from answers with a translation model (Trans), and (4) the combination of them (All). The ranking models are (1) linear combination trained with CA [11], (2) neural ranking model trained with ListNet [2], and (3) ensemble trees trained with LightGBM [8].

For RQ1, we fixed the ranking model to CA and examined four feature sets. We observed the order in the amount of gained credit as follows: CA+BM25F > CA+Trans > CA+Basic > CA+All in the first stage, and CA+BM25F > CA+Trans > CA+Basic in the second stage of the online evaluations. The difference between CA+BM25F and CA+Basic was statistically significant ($p < 0.05$). Under the limited trial efforts such that the submission is allowed only once in a day, reasonably designed feature sets perform similarly but combining all features cause degradation.

For RQ2, we fixed the feature set to Basic and examined three ranking models. We observed the order in the amount of gained credit as follows: ListNet+Basic > CA+Basic > LightGBM+Basic in the first stage and List-Net+Basic > CA+Basic in the second stage of the online evaluations. The difference between ListNet+Basic and CA+Basic in the second stage was statistically significant ($p < 0.05$). The result that ListNet+Basic even outperforms CA+BM25F and CA+Trans (statistically significant in second-stage winning percentage) suggests the further importance of the choice of the ranking model.

References

1. Burges, C.J., Ragno, R., Le, Q.V.: Learning to rank with nonsmooth cost functions. In: Schölkopf, B., Platt, J.C., Hoffman, T. (eds.) Advances in Neural Information Processing Systems 19, pp. 193–200. MIT Press (2007)

2. Cao, Z., Qin, T., Liu, T.Y., Tsai, M.F., Li, H.: Learning to rank: from pairwise approach to listwise approach. In: Proceedings of the 24th International Conference on Machine Learning, ICML 2007, pp. 129–136. ACM, New York(2007)

3. Chapelle, O., Metlzer, D., Zhang, Y., Grinspan, P.: Expected reciprocal rank for graded relevance. In: Proceedings of the 18th ACM Conference on Information and Knowledge Management, CIKM 2009, pp. 621–630. ACM, New York (2009)

4. Chen, M., Li, L., Sun, Y., Zhang, J.: Erler at the NTCIR-13 OpenLiveQ task. In: Proceedings of the 13th NTCIR Conference on Evaluation of Information Access Technologies (2017)

5. Kato, M.P., Manabe, T., Fujita, S., Nishida, A., Yamamoto, T.: Challenges of multileaved comparison in practice: lessons from NTCIR-13 OpenLiveQ task. In: Proceedings of the 27th ACM International Conference on Information and Knowledge Management, CIKM 2018, pp. 1515–1518. ACM, New York (2018)

6. Kato, M.P., Yamamoto, T., Manabe, T., Nishida, A., Fujita, S.: Overview of the NTCIR-13 OpenLiveQ task. In: Proceedings of the 13th NTCIR Conference on Evaluation of Information Access Technologies (2017)

7. Kato, M.P., Yamamoto, T., Manabe, T., Nishida, A., Fujita, S.: Overview of the NTCIR-14 OpenLiveQ-2 task. In: Proceedings of the 14th NTCIR Conference on Evaluation of Information Access Technologies (2019)

8. Ke, G., et al.: LightGBM: a highly efficient gradient boosting decision tree. In: Guyon, I., et al. (eds.) Advances in Neural Information Processing Systems 30, pp. 3146–3154. Curran Associates, Inc. (2017)

9. Kingma, D.P., Ba, J.: Adam: a method for stochastic optimization. CoRR abs/1412.6980 (2014)

10. Manabe, T., Nishida, A., Fujita, S.: YJRS at the NTCIR-13 OpenLiveQ task. In: Proceedings of the 13th NTCIR Conference on Evaluation of Information Access Technologies (2017)

11. Metzler, D., Bruce Croft, W.: Linear feature-based models for information retrieval. Inf. Retrieval **10**(3), 257–274 (2007)

12. Och, F.J., Ney, H.: Improved statistical alignment models. In: Proceedings of the 38th Annual Meeting on Association for Computational Linguistics, ACL 2000, pp. 440–447. Association for Computational Linguistics, Stroudsburg (2000)

13. Oosterhuis, H., de Rijke, M.: Sensitive and scalable online evaluation with theoretical guarantees. In: Proceedings of the 2017 ACM on Conference on Information and Knowledge Management, CIKM 2017, pp. 77–86. ACM, New York (2017)
14. Qin, T., Liu, T.Y., Xu, J., Li, H.: LETOR: a benchmark collection for research on learning to rank for information retrieval. Inf. Retrieval **13**(4), 346–374 (2010)
15. Robertson, S., Zaragoza, H., Taylor, M.: Simple BM25 extension to multiple weighted fields. In: Proceedings of the Thirteenth ACM International Conference on Information and Knowledge Management, CIKM 2004, pp. 42–49. ACM, New York (2004)

Studying Online and Offline Evaluation Measures: A Case Study Based on the NTCIR-14 OpenLiveQ-2 Task

Piyush Arora[✉] and Gareth J. F. Jones

ADAPT Centre, School of Computing, Dublin City University, Dublin 9, Ireland
{Piyush.Arora,Gareth.Jones}@dcu.ie

Abstract. We describe our participation in the NTCIR-14 OpenLiveQ-2 task and our post-submission investigations. For a given query and a set of questions with their answers, participants in the OpenLiveQ task were required to return a ranked list of questions that potentially match and satisfy the user's query effectively. In this paper we focus on two main investigations: (i) Finding effective features which go beyond only-relevance for the task of ranking questions for a given query in Japanese language. (ii) Analyzing the nature and relationship of online and offline evaluation measures. We use the OpenLiveQ-2 dataset for our study. Our first investigation examines user log-based features (e.g number of views, question is solved) and content-based features (BM25 scores, LM scores). Overall, we find that log-based features reflecting the question's popularity, freshness, etc dominate question ranking, rather than content-based features measuring query and question similarity. Our second investigation finds that the offline measures highly correlate among themselves, but that the correlation between different offline and online measures is quite low. We find that the low correlation between online and offline measures is also reflected in discrepancies between the systems' rankings for the OpenLiveQ-2 task, although this depends on the nature and type of the evaluation measures.

Keywords: Learning To Rank models · Question-answer ranking · Online and offline testing · Correlation of online and offline measures

1 Introduction

Interactive websites for community based question answering (CQA) provide opportunities to search and ask questions ranging from critical topics related to health, education and finance to recreational queries for the purpose of fun and enjoyment. Yahoo Chiebukuro (YCH)[1] is a community question answering service which provides a question retrieval system in Japanese language managed by the Yahoo Japan Corporation. The NTCIR-14 OpenLiveQ-2 is a benchmark task which aims to provide an open live test environment using the Yahoo

---

[1] https://chiebukuro.yahoo.co.jp/.

© Springer Nature Switzerland AG 2019
M. P. Kato et al. (Eds.): NTCIR 2019, LNCS 11966, pp. 70–82, 2019.
https://doi.org/10.1007/978-3-030-36805-0_6

Chiebukuro engine where, given a query and a set of questions with their answers, task participants had to return a ranked list of questions. Final evaluation of the results was based on real user feedback. Involving real users in evaluation helps to incorporate the diversity of search intents and relevance criteria by utilising real queries and feedback from users who are engaged in real search tasks, which makes this task more interesting. The submitted systems were evaluated using offline measures such as NDCG@10, ERR@10 and Q-measures and online evaluation metrics using a pairwise preference multileaving approach (discussed later in Sect. 2). This paper describes our participation in the OpenLiveQ-2 task and our post-submission investigations.

Overview of System Submissions: This task focuses on modelling textual based information and click log based information to rank questions to handle the challenges of: (i) queries being ambiguous and having diverse intent, and (ii) modelling user behaviour effectively. A range of Learning To Rank (L2R) models have been investigated in the OpenLiveQ task held at NTCIR-13 and NTCIR-14, respectively [4,7–9]. These L2R models focus on selecting a diverse range of features with effective weights to improve the systems' performance as measured using offline and online evaluation measures. However, apart from [8], not much work has been done in analyzing the nature and type of good features to address the OpenLiveQ task of ranking question-answer pairs for a given query. This observation motivated us to study feature importance for question ranking. In [8], the authors trained a L2R model using a coordinate ascent algorithm for question ranking. To calculate feature importance they removed each feature one at a time, retrained their ranking model and analyzed the relative decrease in the overall scores of NDCG@10, ERR@10 compared to the ranking model learnt using all the features. If a feature is relatively important, its removal led to a greater decrease in the NDCG@10, ERR@10 scores.

As L2R approaches have shown to be quite successful for this task, we also explored L2R models to address the task of ranking question-answer pairs for a given query. We submitted 14 system runs (including the baseline) for the OpenLiveQ-2 task [9]. Our top submission systems were ranked 2 on NDCG@10, ranked 3 on ERR@10, and ranked 6 on Q-measure among the 65 system submissions made to the task. However, these systems which ranked quite high on the offline evaluation measures of NDCG@10, ERR@10 and Q-measure, had a rank below 35 among the 65 submissions made to the task on the online evaluation measure. This contrasting ranking of our system submissions between online and offline evaluation measures motivated us to pursue an investigation on the relationship of the online and offline evaluation measures used in the OpenLiveQ-2 task.

To address the above limitations of finding effective features for ranking questions and to study the nature of online and offline evaluation measures, we set out the following questions for our investigation:

- **RQ-1:** What are the effective features for the task of ranking question-answer pairs for the OpenLiveQ-2 task?

- **RQ-2:** What is the correlation between different online and offline evaluation measures used in the OpenLiveQ-2 task?

Our work seeks to understand the relationship between the online and offline evaluation measures. This topic is of emerging interest in the information retrieval (IR) research community to build better ranking models, and to improve user engagement and satisfaction. The main contributions of our work are:

1. Investigating effective features for the task of question ranking. We find that log-based features reflecting a question's popularity, freshness, etc. dominate the question's ranking over content-based features measuring query-question similarity. Our findings on the OpenLiveQ-2 dataset support the findings of previous work [8] carried out on the OpenLiveQ-1 dataset [4]. In [8] the authors removed one feature at a time to examine the importance of each feature by calculating the decrease in the offline measures (e.g NDCG score) as compared to the equivalent metric score of a combined model built using all the features. In our work we build a comprehensive single model, using all the features, using Gradient Boosting Trees (described later in Sect. 4). We find feature importance by calculating probability estimates of how much the feature contributes to reducing data misclassification. Our work also contributes confirming the claims and findings of previous research on this topic.
2. We study the relationship between different offline and online measures for the OpenLiveQ-2 task. We analyze the fine-grained results output of our 65 system submissions for the task to calculate Pearson correlation between the offline measure scores, such as NDCG, ERR at rank 5, 10, 20 and 50 and Q-measure and the online measure score (described later in Sect. 5). We find that the offline measures correlate highly amongst themselves, but that the correlation between different offline and online measures is quite low. The low correlation between online and offline measures is also reflected in the discrepancies between the systems' rankings for the OpenLiveQ-2 task.

We anticipate that our findings will encourage IR researchers to carefully examine the relationship and variation in system scores and rankings, while using alternative online and offline evaluation measures. The remainder of this paper is organised as follows: Sect. 2 introduces the dataset, tools used and the evaluation strategy of the OpenLiveQ-2 task, Sect. 3 describes an overview of our approach adopted in our participation in this task, Sect. 4 gives results and analysis of our submissions to the task, Sect. 5 descries the relative performance and ranking of the top-k system submissions and describes our investigation studying the relationship between the different evaluation measures used in this task, and finally Sect. 6 concludes.

2   Dataset, Tools and Evaluation

In this section we describe the dataset for OpenLiveQ-2 task, the tools used for this work and the evaluation strategy adopted for the task. As a part of the dataset for the OpenLiveQ-2 task, the organisers provided the query logs and for each query a corresponding set of questions with a best answer retrieved by the YCH engine. Table 1 presents information regarding the number of queries, questions in the training and the test sets. Since the data is in the Japanese language, so as to facilitate participation from diverse and non-native speaking teams in the development of effective systems, the task organisers provided a list of textual features indicating the scores of relevance models such as BestMatch (BM25) [14], Language Model (LM) [13] etc., for a query and a corresponding set of questions. Table 2 presents a list of the complete features which were provided by the task organisers comprising of textual and click-log based information. We refer interested readers to [4, 7] for more details of these features and the dataset construction for this task.

Table 1. Dataset details

Training set	Size	Test set	Size
Number of Queries	1000	Number of Queries	1000
Number of Questions	986125	Number of Questions	985691
Number of click logs	288502	Number of click logs	148388

Table 2. All extracted features provided in the dataset

Title	Id	Snippet	Id	Question body	Id	Best answer	Id	Click logs	Id
tf_sum	F1	tf_sum	F18	tf_sum	F35	tf_sum	F52	answer_num	F69
log_tf_sum	F2	log_tf_sum	F19	log_tf_sum	F36	log_tf_sum	F53	log_answer_num	F70
norm_tf_sum	F3	norm_tf_sum	F20	norm_tf_sum	F37	norm_tf_sum	F54	view_num	F71
log_norm_tf_sum	F4	log_norm_tf_sum	F21	log_norm_tf_sum	F38	log_norm_tf_sum	F55	log_view_num	F72
idf_sum	F5	idf_sum	F22	idf_sum	F39	idf_sum	F56	is_open	F73
log_idf_sum	F6	log_idf_sum	F23	log_idf_sum	F40	log_idf_sum	F57	is_vote	F74
icf_sum	F7	icf_sum	F24	icf_sum	F41	icf_sum	F58	is_solved	F75
log_tfidf_sum	F8	log_tfidf_sum	F25	log_tfidf_sum	F42	log_tfidf_sum	F59	rank	F76
tfidf_sum	F9	tfidf_sum	F26	tfidf_sum	F43	tfidf_sum	F60	updated_at	F77
tf_in_idf_sum	F10	tf_in_idf_sum	F27	tf_in_idf_sum	F44	tf_in_idf_sum	F61		
bm25	F11	bm25	F28	bm25	F45	bm25	F62		
log_bm25	F12	log_bm25	F29	log_bm25	F46	log_bm25	F63		
lm_dir	F13	lm_dir	F30	lm_dir	F47	lm_dir	F64		
lm_jm	F14	lm_jm	F31	lm_jm	F48	lm_jm	F65		
lm_abs	F15	lm_abs	F32	lm_abs	F49	lm_abs	F66		
dlen	F16	dlen	F33	dlen	F50	dlen	F67		
log_dlen	F17	log_dlen	F34	log_dlen	F51	log_dlen	F68		

As outlined in Sect. 1, the OpenLiveQ-2 task had offline and online evaluation phases.

- **Offline evaluation phase**: system performance was measured using *NDCG* [3], *ERR* [1], and *Q-measure* [15, 16].

- **Online evaluation phase**: a *pairwise preference multileaving (ppm)* app-roach was used [6,12] to measure system performance.

The evaluation methodology in OpenLiveQ-2 focused on a two phase online evaluation strategy. In the first phase all the systems were evaluated online to identify the top-k systems, these top-k systems were then compared in detail to ensure that the top systems could be statistically distinguished. For each of the submitted rankings of questions, a multileaving approach was used to form a new set of combined rankings and shown to the users as part of the YCH engine. For a given query each of the questions in the original ranked list that was clicked when presented to a user received a credit, these credit scores were aggregated over the ranked list, and are referred to as the cumulative gain (CG). This CG score was used to rank the systems in the online evaluation phase [4]. To find the top-k systems, the task organisers used a pairwise preference multileaving (PPM) approach which infers pairwise preferences between documents from clicks. The PPM model is based on the assumption that a clicked document is preferred to: (a) all of the unclicked documents above it; (b) the next unclicked document. These assumptions are commonly used in pairwise Learning To Rank models, for more details refer to [12].

3   System Development: Approach Used

In this section we present an overview of our approach to the OpenLiveQ-2 task. Submissions to the previous OpenLiveQ-1 task showed positive results using L2R models [8], thus as a part of our investigation we focused on exploring L2R models [10,11] to rank a set of question-answer pairs given an input query. In L2R models, a ranking function is created using the training data, such that the model can precisely predict the ranked lists in the training data. Given a new query, the ranking function is used to create a ranked list for the documents associated with the query. The focus of L2R technologies is to successfully leverage multiple features for ranking, and to learn automatically the optimal way to combine these features. In this work, we used the Lemur RankLib toolkit [2]. This toolkit provides an implementation of a range of L2R algorithms which have been shown to be successful in earlier work.

Table 3. Feature set

Type of features	Feature's id range	Information type
Title based textual features (Title set)	[F1–F17]	Content based information
Snippet based textual features (Snippet set)	[F18–F34]	Content based information
Question body based textual features (Body set)	[F35–F51]	Content based information
Body answer based textual features (Answer set)	[F52–F68]	Content based information
Click log features (Click set)	[F69–F77]	Logs based information

The main focus of this task was to effectively combine text-based features measuring the similarity of queries with a set of questions and click-based information captured through user logs. We investigated feature selection extensively to determine a good set of features to rank the questions effectively for a given set of test queries. A complete set of features is shown in Table 2. To select features and combine them effectively, we broadly categorised the set of 77 features into 5 main categories, as shown in Table 3. We have diverse feature sets capturing relevance of: (i) user query to question title (*Title set*), (ii) user query to question body (*Body set*), (iii) user query to question snippets (*Snippet set*), (iv) user query to the best answer (*Answer set*), and (v) click logs based information (*Click set*). We explored alternative combinations of these diverse features set.

Run Submissions: As described above, we used the RankLib toolkit for our experiments. Models were trained on the training dataset comprising of about 1M questions (data points) and among which about 300k questions (data points) had information about user interactions. The models were optimised based on the ERR@10 metric. We submitted 14 systems as a part of this investigation. For more details on our approach and different runs that were submitted for this task kindly refer to our system submission paper [9].

Table 4. Offline evaluation scores and system rankings for our submissions for the OpenLiveQ-2 task. The best scores are in boldface.

Systems	System - id	System scores			System ranking		
		NDCG@10	ERR@10	Q-Measure	NDCG-Rank	ERR-Rank	Q-Rank
Best scores		0.333	0.209	0.502	1	1	1
Average scores		0.204	0.128	0.436	NA	NA	NA
System-1	99	0.074	0.044	0.382	59	59	56
System-2	106	0.237	0.171	0.454	32	24	26
System-3	110	0.239	0.138	0.444	31	33	29
System-4	112	0.188	0.137	0.370	36	35	64
System-5	118	0.117	0.106	0.340	45	38	65
System-6	123	0.326	0.202	**0.495**	5	5	**6**
System-7	126	0.204	0.138	0.438	34	34	32
System-8	128	0.285	0.191	0.459	22	20	24
System-9	130	**0.331**	**0.203**	0.464	**2**	**3**	21
System-10	133	0.227	0.148	0.449	33	32	27
System-11	143	0.302	0.189	0.445	19	21	28
System-12	147	0.287	0.179	0.466	21	23	20
System-13	150	0.295	0.181	0.464	20	22	22
System-14	152	0.258	0.154	0.491	25	31	17

4   Results and Analysis

In this section we give results and analysis of our submissions to the OpenLiveQ-2 task. Tables 4 and 5 present the results of our submitted systems for the offline and online evaluation measures respectively. As a part of the official metrics, the organisers reported and compared the ranks and scores of the systems across all three measures NDCG (normalized discounted cumulative gain), ERR (expected reciprocal rank), and Q-measure. As shown in Table 4, we can see some quite distinct variations across the three scores (NDCG@10, ERR@10 and Q-scores) for the system submissions, indicating that these three evaluation metrics do not show consistent trends. For example, System-14 shows Q-scores similar to the best scores of System-6, however the NDCG@10 and ERR@10 scores are quite low compared to the highest scores of System-9.

Table 5. Online evaluation scores and system rankings for our submissions for the OpenLiveQ-2 task. The best two systems are in boldface.

Systems	System - id	Phase-1 online evaluation		Phase-2 online evaluation	
		Cumulative gain	Rank	Cumulative gain	Rank
Best scores		2633.202	1	1867.440	1
Average scores		−4.92	NA	−13.94	NA
System-1	99	−1420.907	61	NA	NA
System-2	**106**	**1843.815**	**7**	**1002.855**	**7**
System-3	110	190.395	40	NA	NA
System-4	112	1721.431	8	428.370	10
System-5	**118**	**2006.333**	**4**	**1129.577**	**6**
System-6	123	70.797	43	NA	NA
System-7	126	1326.385	14	241.362	12
System-8	128	−83.103	46	NA	NA
System-9	130	282.391	38	NA	NA
System-10	133	−40.834	44	NA	NA
System-11	143	171.302	41	NA	NA
System-12	147	452.896	29	−418.210	23
System-13	150	276.774	39	NA	NA
System-14	152	369.791	35	NA	NA

As described in Sect. 2, the online evaluation was conducted in two phases. In the first phase all 61 distinct system submissions were compared in an online setting using a pairwise preference multileaving approach to select the top 30 submissions which were then compared extensively. Table 5 presents results of both the online evaluation phases. In the first phase of online evaluation only two of our submissions (System: ID-128 and ID-133) scored below the average score, the remaining 11 systems performed better than the average score, and five of our thirteen systems were selected to be compared in the final phase of online evaluation. In the final phase of online evaluation only one of our five systems scored below the average score. Our best systems in the online phase

System-5 (ID: 118) and System-2 (ID: 106) were ranked "6" and "7", among the top 30 systems.

Table 6. Top two systems, L2R models were trained using coordinate ascent algorithm with default parameters: tolerance $= 0.001$, iterations $= 25$ and random restarts $= 5$

Systems	System - id	Features combined	Feature set
System-2	106	All 77 Features (F1: F77)	Title+Snippet+Body+Answer+Click
System-5	118	Feature F69:F77 (Click features)	Only click set

To find effective features for the task of ranking question-answer pairs for a given query, we inspect our two best system submissions, System-5 and System-2, in detail. We select these two systems as they perform best in the online evaluation and are the only systems which capture all nine user log based features, as shown in Table 6. We calculated feature importance to find effective features for ranking question-answer pairs for a given query for both System-2 and System-5. We learnt a gradient boosting classification algorithm on the training data using scikit-learn[2]. A gradient boosting algorithm builds a decision tree model using a cross entropy loss function, where node of the trees are the features in the training model. The decision tree model splits the tree node by calculating the gini impurity over all the features. Gini impurity is a measurement of the likelihood of an incorrect classification, it gives a probability estimate of how well the feature splits the data to minimize the data misclassification [17]. The importance of each feature is calculated based on its contribution to splitting the data effectively to perform better classification over the training corpus.

Table 7. Feature rankings representing important features for System-5.

Rank	Feature-id	Feature-name	Important value
1	Feature-71	Number of views	0.246
2	Feature-72	Log of number of views	0.239
3	Feature-77	Updated date	0.226
4	Feature-76	Best rank	0.173
5	Feature-69	Number of answers	0.052
6	Feature-70	Log of number of answers	0.051
7	Feature-75	Is solved	0.010
8	Feature-74	Is voted	0.004
9	Feature-73	Is open	0.000

Table 7 presents features in descending order of importance for predicting effective ranking of questions for System-5. Among the user log based information, features such as number of views, updated date and best rank are relatively

---

[2] https://scikit-learn.org/stable/.

important, indicating that the online results ranking and clicks are influenced by the questions' popularity, freshness, and the relative position in the ranked list, also called position bias [5].

Table 8. Feature rankings representing important features for System-2. Only features greater than 0.01 importance value are shown.

Rank	Feature-id	Feature-name	Value
1	F-71	Number of views	0.116
2	F-70	Log of number of answers	0.114
3	F-75	Is solved	0.058
4	F-76	Best rank	0.043
5	F-6	Log of Idf sum with title	0.034
6	F-40	Log of Idf sum with question body	0.031
7	F-5	Idf sum with Title	0.030
8	F-69	Number of answer	0.026
9	F-68	Log of best answer length	0.025
10	F-41	ICF sum of question body	0.024
11	F-4	Log of norm of TF sum	0.023
12	F-43	Tf-Idf sum of question body	0.023
13	F-39	Idf sum of question body	0.023
14	F-8	Log of Tf-Idf sum of title	0.022
15	F-7	ICF sum of title	0.021
16	F-42	Log of TF-Idf sum of question body	0.021
17	F-35	TF sum of question body	0.019
18	F-56	Idf sum of best answer	0.017
19	F-38	Log of norm of Tf sum of question body	0.015
20	F-57	Log of Idf sum of best answer	0.014
21	F-58	Icf sum of best answer	0.013
22	F-46	Log of BM25 of question body	0.011
23	F-12	Log of BM25 of question title	0.010
24	F-55	Log of norm of TF sum of best answer	0.010

Table 8 presents features in descending order of importance for predicting effective ranking for question-answer pairs for System-2. Most dominant features are "number of views" and "log of number of answer", indicating the popularity of the question. Important features corresponding to content-based information are "query" and "question title" matching followed by "query" and "question body" matching. It is peculiar to see BM-25 scores are ranked 22 and 23, and thus are not as effective relatively in ranking questions for System-2. Similar findings were observed in [8], where the authors found that the top 2 features for question ranking are "log of number of views" and "log of number of answers" indicating the popularity of a question for the NTCIR-13 OpenLiveQ dataset [4]. They found that BM25 scores ranked 15 and 20 in terms of the feature's ranking. However, in their work they found that snippet based features are more effective, while in our work we found that feature matching "query" with "question title" and "question body", respectively is more effective than matching "query" with "snippet".

In summary, we inspected the top features for question ranking by analyzing our top 2 systems for the OpenliveQ-2 task. Some of our findings on the relative importance of features concur with the previous findings reported in [8], thus adding to the reproducibility of the claims with respect to feature importance for question ranking. As most of the traditional IR models work on optimizing relevance to improve over the offline metrics, it becomes necessary to model other aspects such as popularity, diversity and freshness as they tend to perform relatively better on the online metrics. We anticipate that the findings from the feature analysis for the task of ranking questions will encourage more work on understanding how different features correspond to online user behaviour. We have tried to bridge the gap between understanding important features for question ranking and hope that this work will lead to more investigation on the interaction and relationship across these different features.

5   Evaluation Metrics Correlation

In this section we investigate the relationship between the online and offline evaluation metrics. We study how well the online and offline evaluation measures correlate with each other. We use Pearson correlation (r) which is a measure of the linear correlation between two variables x and y as shown in Eq. 1 using scipy library[3].

$$r = \frac{\sum_{i=1}^{n}(x_i - \overline{x})(y_i - \overline{y})}{\sum_{i=1}^{n}(x_i - \overline{x})^2 \sum_{i=1}^{n}(y_i - \overline{y})^2} \tag{1}$$

where n is the sample size, x_i, y_i are the individual sample points indexed with i, $\overline{x} = \frac{1}{n}\sum_{i=1}^{n} x_i$ and $\overline{y} = \frac{1}{n}\sum_{i=1}^{n} y_i$.

As indicated in Sect. 1, there were 65 system submissions made for the OpenLiveQ-2 task. We calculated fine grained evaluation scores for the relevance measures such as NDCG and ERR at different ranks 5, 10, 20 and 50 and Q-Measure for all the 65 systems. We also had the online cumulative gain scores for the online evaluation phase-1 for all 65 systems. We used these 65 data points to find correlation values across different offline and online evaluation measures.

Table 9 presents Pearson correlation results between diverse set of NDCG, ERR at rank 5, 10, 20 and 50 values and for Q-measure and online cumulative gain metrics. The results indicate that the correlation coefficient of the

---

[3] https://docs.scipy.org/doc/scipy/reference/generated/scipy.stats.pearsonr.html.

Table 9. Pearson correlation of all the reported evaluation measures used for the OpenLiveQ-2 task. * and ** indicates that the p-value is more than 0.05 and 0.01 respectively. For all other correlation values p-value is less than 0.01. ≡ indicates that it is a symmetrical relationship. N stands for NDCG, E stands for ERR and CG stands for cumulative gain measures.

Pearson	N@5	N@10	N@20	N@50	E@5	E@10	E@20	E@50	Q	CG
N@5	–	≡	≡	≡	≡	≡	≡	≡	≡	≡
N@10	0.997	–	≡	≡	≡	≡	≡	≡	≡	≡
N@20	0.989	0.997	–	≡	≡	≡	≡	≡	≡	≡
N@50	0.972	0.986	0.995	–	≡	≡	≡	≡	≡	≡
E@5	0.995	0.987	0.976	0.953	–	≡	≡	≡	≡	≡
E@10	0.996	0.992	0.982	0.966	0.999	–	1.00	≡	≡	≡
E@20	0.997	0.994	0.986	0.968	0.997	1.00	–	≡	≡	≡
E@50	0.996	0.994	0.986	0.970	0.996	0.999	1.00	–	≡	≡
Q	0.917	0.939	0.954	0.972	0.892	0.904	0.91	0.912	–	≡
CG	0.333	0.325	0.301**	0.278**	0.368	0.375	0.372	0.374	0.225*	–

offline evaluation metrics is quite high ($r >= 0.9$). However there are some noticeable differences, NDCG@k and ERR@k measures have higher correlation as compared to between NDCG@k and Q-measure and between ERR@k and Q-measure, respectively. Overall, the online evaluation metric CG shows low correlation with the offline evaluation metrics ($r \in [0.225 - 0.375]$). The CG evaluation measure shows higher correlation with ERR@k values as compared with NDCG@k and Q-measures.[4]

The low correlation values between the online and offline evaluation measures explains why the system rankings are quite varied depending on the choice of evaluation metric, as shown in Tables 4 and 5. The online and offline metrics do not go hand in hand and focus on optimization of different aspects and lead to a difference in system ranking. The trained models are tuned and optimized on metrics including NDCG@10 and ERR@10. Thus, for the test queries, question rankings perform quite well when measured using NDCG@10, ERR@10, but evaluating the systems using online metrics, such as cumulative gain, produces low results. For future tasks, involving online and offline evaluation, we recommend the exploration of alternative offline measures for model training and system evaluation that correlates well with the online metrics.

6   Conclusions

In this study we examined the features that are important for question ranking for the OpenLiveQ-2 task. We explored different features to find those that

---

[4] Similar pattern of results were observed using Spearman's and Kendall's Tau correlation metrics during our investigation, results have been omitted because of the space constraints.

contribute effectively for the task of question ranking. We found that features indicating the popularity, freshness and relative position of a question are among the top features for question ranking. Some of these results concur with previous findings on the earlier OpenLiveQ-1 task. Most of IR approaches focus on improving relevance and optimizing models on NDCG, ERR, but we find that in an online setting, there are more diverse features which are important, thus there is a need to incorporate features beyond relevance that capture information effectively. We anticipate the findings in this work will lead to more investigation of the interaction between different features used for ranking questions.

We studied the relationship between online and offline evaluation measures. We calculated Pearson correlation between different offline evaluation measures such as NDCG, ERR at rank 5, 10, 20 and 50 and Q-measure and the online evaluation metric measured using cumulative gain. We found that all the offline evaluation measures correlate well with each other, however the correlation of the offline and online measures is quite low. The low correlation between the online and offline evaluation metrics lead to variation in the ranking of systems depending on the choice of evaluation metric. We anticipate the findings in this work will draw attention from the community, and lead to more work in understanding the relationship between online and offline evaluation measures.

Acknowledgement. This research is supported by Science Foundation Ireland (SFI) as a part of the ADAPT Centre at Dublin City University (Grant No: 12/CE/I2267).

References

1. Chapelle, O., Metlzer, D., Zhang, Y., Grinspan, P.: Expected reciprocal rank for graded relevance. In: Proceedings of the 18th ACM Conference on Information and Knowledge Management, CIKM, pp. 621–630 (2009)
2. Dang, V.: The Lemur Project-Wiki-Ranklib (2013). http://sourceforge.net/p/lemur/wiki/RankLib
3. Järvelin, K., Kekäläinen, J.: Cumulated gain-based evaluation of IR techniques. ACM Trans. Inf. Syst. (TOIS) **20**(4), 422–446 (2002)
4. Kato, M.P., Liu, Y.: Overview of NTCIR-13. In: Proceedings of the 13th NTCIR Conference on Evaluation of Information Access Technologies (2017)
5. Joachims, T., Granka, L.A., Pan, B., Hembrooke, H., Gay, G.: Accurately interpreting clickthrough data as implicit feedback. In: Proceedings of the 28th Annual International ACM SIGIR Conference on Research and Development in Information Retrieval, pp. 154–161. SIGIR (2005)
6. Kato, M.P., Manabe, T., Fujita, S., Nishida, A., Yamamoto, T.: Challenges of multileaved comparison in practice: lessons from NTCIR-13 OpenLiveQ Task. In: Proceedings of the 27th ACM International Conference on Information and Knowledge Management, CIKM, pp. 1515–1518 (2018)
7. Kato, M.P., Nishida, A., Manabe, T., Fujita, S., Yamamoto, T.: Overview of the NTCIR-14 OpenLiveQ-2 task. In: Proceedings of the 14th NTCIR Conference on Evaluation of Information Access Technologies (2019)
8. Manabe, T., Nishida, A., Fujita, S.: YJRS at the NTCIR-13 OpenLiveQ task. In: Proceedings of the 13th NTCIR Conference on Evaluation of Information Access Technologies (2017)

9. Arora, P., Jones, G.J.F.: DCU at the NTCIR-14 OpenLiveQ-2 task. In: Proceedings of the 14th NTCIR Conference on Evaluation of Information Access Technologies (2019)
10. Qin, T., Liu, T.Y., Xu, J., Li, H.: LETOR: a benchmark collection for research on learning to rank for information retrieval. J. Inf. Retrieval **13**(4), 346–374 (2010)
11. Metzler, D., Croft, W.B.: Linear feature-based models for information retrieval. J. Inf. Retrieval **10**(3), 257–274 (2007)
12. Oosterhuis, H., de Rijke, M.: Sensitive and scalable online evaluation with theoretical guarantees. In: Proceedings of the 2017 ACM on Conference on Information and Knowledge Management, CIKM, pp. 77–86 (2017)
13. Ponte, J.M., Croft, W.B.: A language modeling approach to information retrieval. In: Proceedings of the 21st Annual International ACM SIGIR Conference on Research and Development in Information Retrieval, pp. 275–281. SIGIR (1998)
14. Robertson, S., Walker, S., Jones, S., Hancock-Beaulieu, M., Gatford, M.: Okapi at TREC-3. In: NIST Special Publication, no. 500225, pp. 109–123 (1995)
15. Sakai, T.: Evaluating evaluation metrics based on the bootstrap. In: Proceedings of the 29th Annual International ACM SIGIR Conference on Research and Development in Information Retrieval, pp. 525–532. SIGIR (2006)
16. Sakai, T.: On the reliability of information retrieval metrics based on graded relevance. J. Inf. Process. Manag. **43**(2), 531–548 (2007)
17. Breiman, L.: Some properties of splitting criteria. J. Mach. Learn. **24**(1), 41–47 (1996)

QA Lab for Political Information

Cue-Phrase-Based Text Segmentation and Optimal Segment Concatenation for the NTCIR-14 QA Lab-PoliInfo Task

Katsumi Kanasaki$^{(\boxtimes)}$, Jiawei Yong$^{(\boxtimes)}$, Shintaro Kawamura, Shoichi Naitoh, and Kiyohiko Shinomiya

Ricoh Company, Ltd., Ebina, Japan
{katsumi.kanasaki,kai.yuu,shintaro.kawamura,
shohichi.naitoh,kshino}@jp.ricoh.com

Abstract. The segmentation subtask of the NTCIR-14 QA Lab-PoliInfo task is finding a segment of text in assembly minutes that corresponds to a summary sentence. We divided the segmentation subtask into two steps, segmentation and search. Cue phrases were effectively used to detect segment boundaries. We compared five methods for detecting segment boundaries: a rule-based method, three supervised learning methods, and a novel semi-supervised learning method. The supervised models were trained using minutes data (in Japanese) we had segmented. In the search step, contiguous segments were concatenated to form larger segments, and the segment that maximized the value of a formula was selected as the answer. We compared the proposed formula with the conventional BM25 formula. We achieved the highest F-measure during the NTCIR-14 formal run despite our method's simplicity.

Keywords: Text segmentation · Semi-supervised method · Hierarchical attention network · Okapi BM25

1 Introduction

Identifying the source information from citations in publications or on the Internet is important to validate cited contents. The segmentation subtask of the NTCIR-14 QA Lab-PoliInfo task [9] is finding a sequence of utterances that corresponds to a given summary sentence. The utterances are taken from assembly minutes, which are a record of the activities of a governmental assembly.

It is not necessary to topically segment the utterances to find the correspondence, as demonstrated by Yokote and Iwayama [18], who solved the PoliInfo problem without using topical segmentation. A speaker's speech is a sequence of utterances, and all subsequences are candidate segments. However, if only part of a sequence on a topic is chosen, the contents can be misunderstood. This is why we consider the task to consist of a segmentation step and a search step. The segmentation step splits all the utterances into segments, each of which

© Springer Nature Switzerland AG 2019
M. P. Kato et al. (Eds.): NTCIR 2019, LNCS 11966, pp. 85–96, 2019.
https://doi.org/10.1007/978-3-030-36805-0_7

deals with a topic. The search step finds a segment or a concatenated segment that corresponds to the summary.

Both text segmentation and text search have been widely studied.

TextTiling [7] and C99 [4] are well known unsupervised text segmentation methods. They rely on lexical cohesion throughout a segment. It is difficult to precisely locate the boundary between segments. Although cue phrases, which often appear in assembly utterances, are effective in locating the boundaries, TextTiling and C99 do not consider them. Eisenstein and Barzilay [6] presented an unsupervised method, in which different segments are assumed to be derived from different language models, and texts around segment boundaries are derived from another language model that reflects cue phrases. Although this approach considers cue phrases in an unsupervised method, their effects in the experiment were negligible.

Cue phrases were also used by Terazawa et al. [15] and Kimura et al. [8] in the PoliInfo segmentation subtask. Although they used a rule-based approach to find cue phrases, cue phrases can play important roles in supervised segmentation as well [3], and supervised approaches that use neural networks have recently been reported [2,10]. A large dataset like Wikipedia is often used to train a model. It is, however, difficult to directly apply a model to assembly minutes because assembly utterances have unique characteristics. We thus prepared our own training data: assembly minutes (in Japanese) with tags indicating segment boundaries.

The segment search task is similar to a document search task, and the widely used TF-IDF scheme can be applied. However, when the given summary contains more than one topic, a segment consisting of smaller segments must be found. Segments can be, in this way, concatenated during the search, which means the length of a segment should be taken into account. Using Okapi BM25 formula [13] is one way to prioritize shorter documents.

Semantic textual similarity represents the degree of semantic relevance between two snippets of text [1], and pseudo relevance feedback [11] is an unsupervised approach to incorporating the semantic similarity into the search task. Since in our case the words in the summary are simply the same as those in the utterances, there is no compelling reason to consider semantic similarity.

Eisenstein [5] presented hierarchical text segmentation. In our approach, we consider three levels in a hierarchical structure of segments. Each speech segment is coarse-grained and consists of medium-grained segments, each dealing with similar topics. These medium-grained segments are typically a concatenation of fine-grained segments, each dealing with a single topic. Each fine-grained segment is a sequence of utterances, and the first and the last utterances of the segment often contain cue phrases. Our aim is to find a medium-grained segment that is close to the given summary.

2   Methods

In our approach, the PoliInfo segmentation subtask is divided into two steps, the segmentation itself and segment search. In the segmentation step, a set of assembly minutes is received, and the utterances are split into a series of topical segments. In the search step, a summary sentence together with the date of the assembly and the speaker identification is received, and a segment or a sequence of segments corresponding to the summary is found.

Assembly utterances contain many cue phrases, which listeners often rely upon to recognize the topic boundaries. We compared five methods that use cue phrases to find these boundaries.

In the search step, we found that the received summary sometimes corresponds to a sequence of more than one segment identified in the previous step. The speech is a sequence of fine-grained segments, and all the subsequences of the sequence are candidates for the segment corresponding to the summary. The subsequence that maximize the value of a formula is considered to be the best candidate. We compared our original formula with the Okapi BM25 formula.

2.1   Cue-Phrase-Based Segmentation

Cue phrases appearing in the first and last utterance of each topical segment are used in the segmentation step. The problem is modeled as a classification problem, the problem of determining whether each utterance starts a segment.

To train and evaluate the segmentation models, we prepared a dataset of utterances that we annotated by hand. They are split into segments to cover all the utterances of each speaker. The dataset consists of two parts, corresponding to the dates of the assembly minutes. The first part, which was used for training, contained 4804 utterances split into 995 segments. The second part, which was used for development, contained 3438 utterances split into 683 segments.

Five segmentation methods were compared.

1. Rule-based

   During the annotation, we found many cue phrases. The regular expressions used to find the phrases are shown in Table 1. An utterance was classified as the first utterance in a segment if and only if the text of the utterance matched the opening pattern or the text of the previous utterance matched the closing pattern but not the opening pattern.

2. Support vector machine (SVM)

   An SVM classifier was learned from the prepared training dataset.

   Each utterance was first broken into words using the MeCab Japanese language morphological analyzer and the IPAdic lexicon. The lemmas of the ten words at the beginning of each utterance and the ten words at the end of each utterance were extracted and converted into a bag-of-words (BoW) vector. The BoW vectors were then compressed into 100-dimensional vectors with latent semantic indexing. For each utterance, the vector of the utterance and that of the previous utterance were concatenated. The concatenated 200-dimensional vectors were used as feature vectors for the SVM classifier.

3. Long short-term memory (LSTM)

Table 1. Regular expressions used to find cue phrases.

pattern	regular expression
opening	^まず|^最初に|^初めに|^次に|^次いで|^最後に|^終わりに
	|^[一二三四五六七八九十]+点目
	|^[^、]+について(で(す|あります|ございます)(が|けれど)
	|^終わり(ま|で)す。|^以上で|^ありがとうございま
	|他の質問に(ついて|つきまして)は
closing	伺い[^、]*ます。|お尋ね[^、]*します。|お答えください。
	|(見解|所見|答弁)を求め[^、]*ます。
	|(いかがで|どうで)(しょうか|すか)。
	|.+質問を(終わります|終了します)。

An LSTM classifier was learned from the prepared training dataset.

For each utterance, the ten words at the beginning of the previous utterance, the ten words at the end of the previous utterance, the ten words at the beginning of the utterance, and the ten words at the end of the utterance were collected. The word embeddings of the 40 words were concatenated and given to a uni-directional LSTM network. The 100-dimensional word2vec model (pre-trained by researchers at Tohoku University) [14] was used. The dropout technique was used to avoid overfitting.

4. Hierarchical attention network (HAN) [17]

A HAN classifier was learned from the prepared training dataset.

For each utterance, the ten words at the beginning of the previous utterance, the ten words at the end of the previous utterance, the ten words at the beginning of the utterance, and the ten words at the end of the utterance were collected. The surface forms of the words were used instead of lemmas. Each sequence of ten words was given to a word-level network consisting of an embedding layer, a bi-directional gated recurrent unit (GRU) layer, and an attention layer. The embedding layer was trainable and randomly initialized. The four results of the word-level network were given to a sentence-level network consisting of a bi-directional GRU and an attention layer.

5. Semi-supervised

Our semi-supervised segmentation method learned from the original minutes data, which contained 84,905 utterances, instead of from the training dataset, which contained the gold standard.

The first utterance of a speaker is also the first utterance of a segment, and we constructed a classifier from the features of such utterances. Applying the classifier to the complete set of minutes, we extracted the candidate first utterances of segments. The previous utterance of each first utterance candidate is a candidate last utterance of a segment. We then constructed another classifier from the features of the previous utterances. Applying this classifier to the complete set of minutes, we extracted the last utterance candidates of segments. The next utterance of each candidate is a candidate first utterance

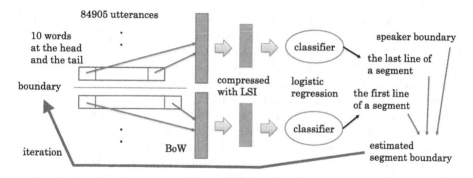

Fig. 1. In our semi-supervised segmentation method, segment boundaries are learned through bootstrapping.

of a segment. We gradually improved the results through this bootstrapping method (see Fig. 1). This iterative process was inspired by the bootstrapping approach to named entity recognition [12].

In our experiment, the BoW vectors were compressed into 200-dimensional vectors. Two classifiers for the first and last utterances were constructed in each iteration. Positive samples were first collected from the boundaries among speakers, and negative samples were randomly collected from around positive samples. Logistic regression was used for the classifier in order to tune the balance between recall and precision. When the estimated probability exceeded a threshold value, the sample was considered to be positive. Each boundary between two utterances was considered to be a boundary between two segments in the next iteration if one of three conditions was met:

- it was a boundary between speakers,
- the utterance immediately after the boundary was classified as a first utterance, or
- the utterance immediately before the boundary was classified as a last utterance.

6. No segmentation

No segmentation was used. This means that each utterance constituted a fine-grained segment.

2.2 Segment Search with Optimal Segment Concatenation

The next step was to find a segment or a sequence of contiguous segments corresponding to each sample of the PoliInfo segmentation subtask. Each sample contained a date, a topic, and a subtopic along with the identification of the two speakers and two summary sentences. A speaker and a summary sentence corresponded to a question, and the other pair corresponded to the answer to the question.

The date and the speaker were first used to find a sequence of segments. The speaker matching was not straightforward because a speaker's name was recorded in the assembly minutes, while the speaker's position was given in the summary dataset. Fortunately, remarks regarding changes in speaker were also recorded in the minutes. We were able to use the remarks to match the speaker because each remark included both the position and the name of the speaker.

A speaker often gave answers to more than one questioner, and the answers sometimes contained similar contents. To distinguish the answers, we chose the sequence of answer segments following each identified question segment.

We then used the summary sentence together with the subtopic to identify corresponding segments. Hereinafter, a summary sentence together with a subtopic is called a summary. The characters in each summary were first converted into full-width characters, and the digits were converted into Chinese characters to match the text in the minutes. The summary was then broken into words using MeCab with IPAdic.

We constructed a simplified probabilistic model to find the sequence of contiguous segments. Only the set of words in the summary was used; the order of the words was ignored. Let us assume that words $t_i (i = 1, \ldots, k)$ in the summary appear in $df(t_i)$ utterances among all N utterances. Provided that words appear independently in utterances, the probability that all the words appear in a sequence of n contiguous utterances is given by

$$P = \prod_{i=1}^{k} \left(1 - \left(1 - \frac{df(t_i)}{N} \right)^n \right) \simeq \prod_{i=1}^{k} n \frac{df(t_i)}{N}. \tag{1}$$

This approximation is based on the assumption that $n\frac{df(t_i)}{N}$ is small. If the function idf is defined as $idf(t_i) = \log(\frac{N}{df(t_i)})$, we get

$$\log(\frac{1}{P}) \simeq \sum_{i=1}^{k} idf(t_i) - \lambda k \log(n), \tag{2}$$

where weight parameter $\lambda = 1$. We search for the sequence of contiguous segments that maximizes the value of the IDF-based formula.

Since words do not independently appear in utterances, we conducted the experiment by changing weight parameter λ. The parameter was tuned so that the highest F-measure was achieved with the training dataset provided by the QA Lab task organizer. The portion of data that corresponds to the training dataset we prepared was excluded because the learning process of three of the methods compared depends on that portion.

If multiple occurrences of a word in a segment are separately counted, the equation becomes

$$\log(\frac{1}{P}) \simeq \sum_{i=1}^{k} tf(t_i) idf(t_i) - \lambda (\sum_{i=1}^{k} tf(t_i)) \log(n), \tag{3}$$

where $tf(t_i)$ is the number of occurrences of the word t_i. We define this equation as a TF-IDF-based formula.

Another possible formula is Okapi BM25:

$$\sum_{i=1}^{k} idf(t_i) \cdot \frac{tf(t_i) \cdot (k_1 + 1)}{tf(t_i) + k_1 \cdot (1 - b + b \cdot \frac{L_d}{L_{ave}})}, \tag{4}$$

where L_d is the length of a document, L_{ave} is the average document length, and k_1 and b are tunable parameters. In our case, a segment or a sequence of contiguous segments was considered to be a document. The formula gives higher priorities to shorter documents if b is not zero.

We used the Anserini toolkit [16] together with the Kuromoji Japanese morphological analysis engine to get the BM25 results. Anserini supports the RM3 pseudo relevance feedback mechanism. We tested both pure BM25 and BM25 combined with RM3.

3   Results

We applied the five segmentation methods to the prepared development dataset. The results are summarized in Table 2. Recall and precision are based on the classification problem that labels the first utterance of a segment as positive. P_k is a metric widely used for segmentation tasks.

Table 2. Performance of five methods in segmentation step. Mean values of ten runs are shown. A smaller value of P_k is better.

Method	Recall	Precision	P_k
Rule-based	0.977	0.928	0.040
SVM	0.938	0.849	0.083
LSTM	0.808	0.918	0.122
HAN	0.968	0.932	0.043
Semi-supervised	0.803	0.789	0.122

The results for the PoliInfo segmentation subtask are shown in Table 3. The test dataset was the gold standard provided by the task organizer and included 83 questions and 83 answers.

Except in the no segmentation case, the IDF-based formula was used with weight λ 0.4 for questions and 0.7 for answers. In the no segmentation case, the TF-IDF-based formula with λ 0.7 for questions and 0.9 for answers gave the best results.

We also used the BM25 formula combined with rule-based segmentation. The highest F-measure for the training dataset was achieved when k_1 was 3.0 and b was 0.9 for questions, and k_1 was 3.5 and b was 1.0 for answers.

We also tried BM25 with RM3 pseudo relevance feedback, but we could not get an F-measure higher than that with pure BM25.

Table 3. Performance of five segmentation methods and three search formulae applied to test dataset. The values that differ from the ones shown in the overview paper of the PoliInfo task [9] are the mean values of ten runs.

Segmentation method	Search formula	Recall	Precision	F1		
		All	All	All	Question	Answer
Rule-based	IDF-based	0.882	0.909	0.896	0.881	0.925
Rule-based	BM25	0.937	0.881	0.908	0.903	0.919
SVM	IDF-based	0.843	0.873	0.858	0.828	0.919
LSTM	IDF-based	0.939	0.708	0.808	0.764	0.917
HAN	IDF-based	0.888	0.885	0.886	0.862	0.938
Semi-supervised	IDF-based	0.859	0.780	0.818	0.792	0.871
No segmentation	TF-IDF-based	0.780	0.746	0.763	0.767	0.751

4 Discussion

Comparing the results in Table 2 with those in Table 3, we see that better segmentation tends to give better results. This means that the segmentation step before the search step is effective.

It is not surprising that the rule-based method had good segmentation performance because the human annotator also had similar patterns for cue phrases in mind. Although the SVM and LSTM methods were not as good as the rule-based method, we think that machine learning methods can produce better (or at least comparable) results than a rule-based method. This is why we tried the HAN method. The HAN method had performance close to that of the rule-based method.

The semi-supervised method is interesting because a large training dataset does not need to be prepared. Additional labeling was unnecessary because we relied on the boundaries between speakers to get the initial labels. Some cue-phrase patterns may not appear in the training data, but the semi-supervised method can find such patterns.

Figure 2 shows how recall and precision in the segmentation step changed with the number of iterations in the semi-supervised method. The plots show that recall and precision converged in the cases shown. When the probability threshold parameter was 0.80, precision decreased with an increase in the number of iterations, which is not preferable. (Iteration zero is exceptional: precision was 1 because boundaries between speakers are always boundaries between segments). Recall is considered to be more important than precision for the segmentation step because segments are concatenated in the search step. We chose a threshold of 0.85 and the 8th iteration for the submission.

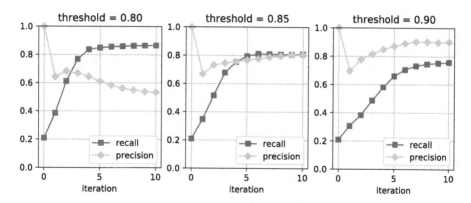

Fig. 2. In our semi-supervised method, convergence through iteration depended on the probability threshold parameter.

Figure 3 shows the effect of weight parameter λ in the search step. Recall, precision, and F-measure for the PoliInfo segmentation subtask are plotted. A smaller λ yielded longer sequences of segments, which means higher recall and lower precision. There is an optimum value of λ to get the highest F-measure. The optimum value for answers differed from that for questions. We think the differences in the length and characteristics of utterances are the reasons for the difference.

The traditional BM25 outperformed our original IDF-based formula on the F-measure for questions. (Although BM25 also outperformed it on recall, the balance between recall and precision depends on the parameters.) This suggests the possibility of better formulae.

Typical values for the BM25 parameters are 1.2 for k_1 and 0.7 for b. In our experiment, the performance was not sensitive to k_1 and was sensitive to b, as shown in Fig. 4, and 0.9 or 1.0 for b is greater than the typical value. This means that selecting concatenated segments must have different characteristics from selecting variable-length documents.

The RM3 pseudo relevance feedback did not improve the results with BM25. In this task, query expansion is ineffective because the terms used in the summary are simply those in the utterances. Query expansion may help in finding assembly utterances from citations in news and blog articles.

When our approach is applied to news and blog articles, it is possible that dates and/or speaker names are not given in the citations. In testing the BM25 formula, the segment with the highest BM25 score among all segments for all dates and speakers matched the specified date and speaker for 76 out of 83 questions and 71 out of 83 answers. This suggests that this task is not very hard even if dates and/or speakers are not given.

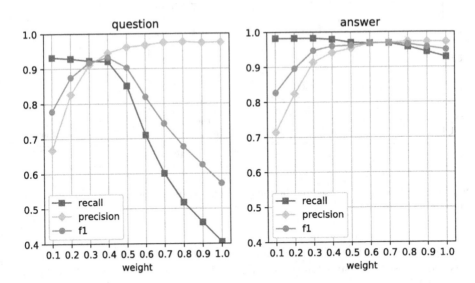

Fig. 3. Balance between recall and precision changed with weight parameter λ. The training dataset provided by the task organizer and rule-based segmentation were used for this plot.

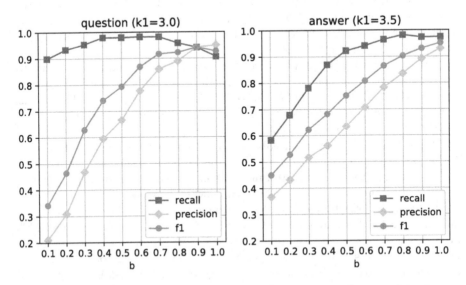

Fig. 4. Parameter b in BM25 must be tuned so that segments of appropriate size are found. The training dataset provided by the task organizer and rule-based segmentation were used for this plot.

5 Conclusion

Our contributions are summarized as follows.

1. We have presented an approach to identifying source information of a citation by concatenating fine-grained speech segments.
2. We showed that assembly utterances can be effectively segmented by detecting cue phrases.
3. Although a rule-based approach results in good segmentation, we showed that a neural network approach can achieve almost the same precision and recall.
4. We also presented a semi-supervised method for segmentation, in which additional labeling is not required.
5. We presented a simple but effective probabilistic model for finding a sequence of segments that corresponds to a given summary.

Our approach achieved the highest F-measure in the NTCIR-14 formal run.

For the PoliInfo segmentation subtask, citation sentences are taken from the newsletter of an assembly. The terminologies used in the newsletter are carefully chosen so that they do not differ from the original utterances. In other citation sources, however, different terminologies may be used. More advanced methods that take syntactic and semantic features into account are thus needed. Our approach is a strong basis for further study.

References

1. Agirre, E., Diab, M., Cer, D., Gonzalez-Agirre, A.: Semeval-2012 task 6: a pilot on semantic textual similarity. In: Proceedings of the First Joint Conference on Lexical and Computational Semantics-Volume 1: Proceedings of the Main Conference and the Shared Task, and Volume 2: Proceedings of the Sixth International Workshop on Semantic Evaluation, pp. 385–393. Association for Computational Linguistics (2012)
2. Badjatiya, P., Kurisinkel, L.J., Gupta, M., Varma, V.: Attention-based neural text segmentation. In: Pasi, G., Piwowarski, B., Azzopardi, L., Hanbury, A. (eds.) ECIR 2018. LNCS, vol. 10772, pp. 180–193. Springer, Cham (2018). https://doi.org/10.1007/978-3-319-76941-7_14
3. Beeferman, D., Berger, A., Lafferty, J.: Statistical models for text segmentation. Mach. Learn. **34**(1–3), 177–210 (1999)
4. Choi, F.Y.Y.: Advances in domain independent linear text segmentation. In: Proceedings of the 1st North American Chapter of the Association for Computational Linguistics Conference, pp. 26–33 (2000)
5. Eisenstein, J.: Hierarchical text segmentation from multi-scale lexical cohesion. In: Proceedings of Human Language Technologies: The 2009 Annual Conference of the North American Chapter of the Association for Computational Linguistics, pp. 353–361. Association for Computational Linguistics (2009)
6. Eisenstein, J., Barzilay, R.: Bayesian unsupervised topic segmentation. In: Proceedings of the Conference on Empirical Methods in Natural Language Processing, pp. 334–343. Association for Computational Linguistics (2008)

7. Hearst, M.A.: Texttiling: segmenting text into multi-paragraph subtopic passages. Comput. linguist. **23**(1), 33–64 (1997)
8. Kimura, T., Tagami, R., Katsuyama, H., Sugimoto, S., Miyamori, H.: KSU systems at the NTCIR-14 QA Lab-PoliInfo task. In: Proceedings of the 14th NTCIR Conference on Evaluation of Information Access Technologies, pp. 251–267 (2019)
9. Kimura, Y., et al.: Overview of the NTCIR-14 QA Lab-PoliInfo task. In: Proceedings of the 14th NTCIR Conference on Evaluation of Information Access Technologies, pp. 121–140 (2019)
10. Koshorek, O., Cohen, A., Mor, N., Rotman, M., Berant, J.: Text segmentation as a supervised learning task. In: NAACL-HLT (2018)
11. Lavrenko, V., Croft, W.B.: Relevance based language models. In: Proceedings of the 24th Annual International ACM SIGIR Conference on Research and Development in Information Retrieval, SIGIR 2001, pp. 120–127. ACM, New York (2001)
12. Riloff, E., Jones, R., et al.: Learning dictionaries for information extraction by multi-level bootstrapping. In: AAAI/IAAI, pp. 474–479 (1999)
13. Robertson, S., Walker, S., Jones, S., Hancock-Beaulieu, M., Gatford, M.: Okapi at TREC-3. In: Proceedings of the Third Text REtrieval Conference (1994)
14. Suzuki, M., Matsuda, K., Sekine, S., Okazaki, N., Inui, K.: Neural joint learning for classifying Wikipedia articles into fine-grained named entity types. In: Proceedings of the 30th Pacific Asia Conference on Language, Information and Computation: Posters, pp. 535–544 (2016)
15. Terazawa, K., Shirato, D., Akiba, T., Masuyama, S.: AKBL at NTCIR-14 QA Lab-PoliInfo task. In: Proceedings of the 14th NTCIR Conference on Evaluation of Information Access Technologies, pp. 190–197 (2019)
16. Yang, P., Fang, H., Lin, J.: Anserini: enabling the use of Lucene for information retrieval research. In: Proceedings of the 40th International ACM SIGIR Conference on Research and Development in Information Retrieval, pp. 1253–1256. ACM (2017)
17. Yang, Z., Yang, D., Dyer, C., He, X., Smola, A., Hovy, E.: Hierarchical attention networks for document classification. In: Proceedings of the 2016 Conference of the North American Chapter of the Association for Computational Linguistics: Human Language Technologies, pp. 1480–1489 (2016)
18. Yokote, K., Iwayama, M.: NAMI question answering system at QA Lab-PoliInfo. In: Proceedings of the 14th NTCIR Conference on Evaluation of Information Access Technologies, pp. 278–288 (2019)

Automatic Training Data Construction and Extractive Supervised Summarization for NTCIR-14 QA Lab-PoliInfo

Satoshi Hiai[✉], Yuka Otani, Takashi Yamamura, and Kazutaka Shimada

Department of Artificial Intelligence, Kyushu Institute of Technology, Fukuoka, Japan
{s_hiai,y_otani,t_yamamura,shimada}@pluto.ai.kyutech.ac.jp

Abstract. On the summarization task at NTCIR-14 QA Lab-PoliInfo, participants of the task need to generate a summary corresponding to an assembly member speech in assembly minutes within the limit length. Our method extracts important sentences to summarize an assembly member speech in the minutes. Our method applies a machine learning model to predict the important sentences. However, the given assembly minutes' data do not contain information about the importance of the sentences. As a result, we cannot directly utilize machine learning techniques for the task. Therefore, we construct training data for the importance prediction model using a word similarity between sentences in a speech and those in the summary. In addition, we apply the sentence reduction process. In the process, we consider characteristics of summaries of assembly minutes to avoid removal of important words in extracted sentences. On the evaluation, all the scores by our supervised method with the constructed data outperformed unsupervised and supervised baseline methods. The result shows the effectiveness of our method.

Keywords: Extractive summarization · Sentence extraction · Automatic dataset construction · Sentence reduction · Machine learning

1 Introduction

Document summarization is a task of automatically generating a summary for a given document. A summarization task conducted at the workshop NTCIR-14 QA Lab-PoliInfo [4]. For the summarization task, an assembly person's speech and a limit length of the summary are given. Participants of the task need to generate a summary corresponding to the speech within the limit length.

Summarization methods are mainly classified into two categories: extractive and abstractive. Abstractive summarization methods can generate words and phrases not contained in the source text with pre-trained vocabulary. On the other hand, extractive summarization methods can generate grammatically well-formed summaries because the methods extract a set of sentences in the source

© Springer Nature Switzerland AG 2019
M. P. Kato et al. (Eds.): NTCIR 2019, LNCS 11966, pp. 97–109, 2019.
https://doi.org/10.1007/978-3-030-36805-0_8

text. Assembly minutes are primarily the evidential record of the assembly activities. The preciseness of the summaries is more important than the readability of those. Therefore, we utilize an extractive summarization method for the task.

There are two types of extraction methods: supervised methods and unsupervised methods. Supervised methods usually show better performance than unsupervised methods. Therefore, we use a supervised method. Our method extracts important sentences to summarize a speech. Our method applies a machine learning approach to predict the importance of sentences in documents. We require labeled data for learning the importance prediction model. However, the assembly minutes' data do not contain the importance labels for sentences. Therefore, we need to assign the importance scores to the sentences in the assembly minutes. We have proposed a method to assign the importance scores automatically [1]. For the assignment, we utilize that the words in the summaries are used in the assembly minutes. We calculated the importance scores using a word similarity. We used the data with the importance scores to train the importance prediction model. The model predicted the importance of each sentence in the assembly minutes. We extracted sentences on the basis of the importance score. In addition, we applied a sentence reduction process. We compressed sentences to generate summaries. We confirmed the effectiveness of our supervised method with the constructed data using the importance assignment process. However, the method often removed important words contained in reference summaries due to the sentence reduction process.

In this study, we apply an automatic importance score assignment process to construct training data and sentence reduction process as with our previous study. In the reduction process, to avoid removal of important words in extracted sentences, we consider characteristics of summaries of assembly minutes such as a position of important words and redundant descriptions. We compress sentences to generate a meaningful summary under a length constraint. We verify the effectiveness of our supervised sentence extraction methods using constructed data and the sentence reduction process.

2   Related Work

In this section, first, we describe the workshop NTCIR-14 QA Lab-PoliInfo. Next, we describe the summarization methods proposed in NTCIR-14 QA Lab-PoliInfo and sentence reduction methods.

2.1   Summarization Task at NTCIR-14 QA Lab-PoliInfo

The NTCIR (NII Testbeds and Community for Information access Research) Workshop is a series of evaluation workshops designed to enhance research in information access technologies. For the summarization task at NTCIR-14 QA Lab-PoliInfo [4], organizers provided an assembly minutes corpus containing assembly member speeches and reference summaries corresponding to the speeches. Table 1 shows an example of a speech and a reference summary in the

Table 1. An example of a speech and a reference summary

Speaker	岡田眞理子 (民主党) (Mariko Okada (Democratic Party of Japan))
Main topic	一体的な築地まちづくり進めよ (Promote the integrated development of Tsukiji city.)
Subtopic	晴海客船ターミナル (Harumi passenger ship terminal)
Reference summary	活性化を促進すべき。(We should promote the activation.)
Speech	晴海の客船ターミナルについて伺います。 (We discuss Harumi passenger ship terminal.) ... 美しさがひときわ目立つ立派な施設ですから、テレビドラマや雑誌のロケ地としてお目にかかることは多いのですが、客船ターミナルとして活性化を促進する 努力をするべき と考えますが、見解を伺います。 (Since the passenger ship terminal is very beautiful, the terminal is often used as a location for TV dramas and magazines. However, we should promote the activation of the terminal as a location for the passenger ship terminal.)

data. The assembly member speeches contain speaker, main topic, and subtopic of the speech. The input of the task is an assembly member speech and a limit length of a summary. The output of the task is a summary corresponding to the speech.

2.2 Summarization Methods in NTCIR-14 QA Lab-PoliInfo

We describe summarization methods proposed in the NTCIR-14 QA Lab-PoliInfo. Kimura et al. [3] have proposed an abstractive summarization method. They generated summaries using a sequence to sequence (seq2seq) model with recurrent neural networks. In general, abstractive summarization methods need a lot of pairs of speeches and reference summaries. However, the assembly minutes corpus did not contain enough pairs. Therefore, several participants have proposed extractive summarization methods.

Extractive summarization methods generally extract a set of sentences in a document to generate summaries. There are two types of extraction methods: supervised methods and unsupervised methods.

Several unsupervised methods have been proposed in the summarization task at NTCIR-14 QA Lab-PoliInfo. Tang et al. [13] have generated summaries with the method based on the TextRank algorithm [8]. TextRank is a typical unsupervised extractive summarization method and is a graph-based ranking method. The basic idea of the method is important sentences are similar to the other important sentences. Terazawa et al. [14] have identified key expressions and extracted sentences containing the expressions. Shinjo et al. [12] have generated summaries using the method based on an optimization problem of selecting important sentences. They calculated sentence weights using the features such

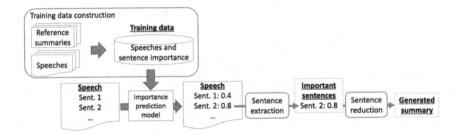

Fig. 1. Overview of our methods.

as a term frequency, words, and parts-of-speech tags and selected the best combination of sentences using the weights.

Two supervised extractive methods have proposed on the summarization task at NTCIR-14 QA Lab-PoliInfo. We have already proposed a method using a sentence importance prediction model [1]. The method constructed training data containing importance scores of each sentence for the model automatically. The method assigned the importance scores to each sentence in a speech on the basis of a word similarity between each sentence in a speech and a summary. The method trained a regression model to predict the importance scores. Ogawa et al. [10] also have constructed training data on the basis of the similarity. They assigned binary labels (important or not) to each sentence. They trained multiple random forest classifiers using the data.

2.3 Sentence Reduction

Sentence reduction is a process to remove redundant words or phrases from the original sentence. In our previous study [1], we supposed that function words in the beginning and end of a sentence were unimportant. We removed the words to compress sentences. However, the methods often removed important words.

Many researchers have proposed sentence reduction methods. Kiyota et al. [5] have generated summaries of web text by trimming a sentence tree representing the dependency between words. By considering the dependency structure, they generated grammatically correct summaries. Mikami et al. [9] have proposed a sentence reduction method to generate summaries of news articles. They considered characteristics of the news article and proposed the summarization method specific to news articles.

In this paper, we consider the characteristics of the assembly minutes data for our sentence reduction process. In the last sentence of the speech in Table 1, the underlined expression in the reference summary appears in the last part of the sentence. In the given corpus in the summarization task at NTCIR-14, an important part of a sentence often appears in the last part of a sentence. Therefore, our method removes preceding phrases in a sentence to generate summaries under a length constraint. For evaluation, we compare our method considering

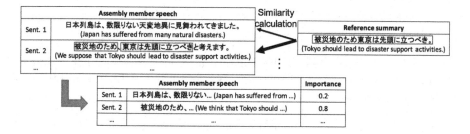

Fig. 2. Example of the training data construction.

the characteristics of the data with the method based on the sentence tree trimming by Kiyota et al.

3 Proposed Method

Figure 1 shows an overview of the proposed method. We construct a sentence importance prediction model. In Sect. 3.1, we explain training data construction for the model. In Sect. 3.2, we explain features for the model. In the sentence extraction process, we extract sentences on the basis of the predicted importance scores and a limit length of a speech. We explain the process in Sect. 3.3. In the next process, if the number of characters contained in extracted sentences is more than the limit length, we apply a sentence reduction process to generate summaries within the limit length. We explain the process in Sect. 3.4.

3.1 Training Data Construction

We explain the training data construction. Figure 2 shows an example of this process. We assign importance scores to each sentence in the assembly minutes. For the assignment, we use a word similarity between the sentences in the speech and the sentences in the summary. We assign the similarity scores to each sentence as importance scores. We use the cosine similarity between bag-of-words representations of sentences as the similarity measure. When a reference summary consists of two or more sentences, the similarity scores corresponding to each sentence in the speech are calculated for all sentences in the summary. Then, we assign the maximum similarity score to each sentence in the speech as the importance score.

3.2 Features

We construct the sentence importance prediction model using training data constructed with a word similarity measure explained in the previous section. We use RBF kernel regression to predict the importance of sentences. The features of the model are as follows.

- Bag-of-words: We use MeCab tokenizer [6] with IPAdic.
- Position: The features of sentence positions were used in some summarization studies [2,11]. If a sentence is the i-th sentence in a speech containing N sentences, we assign $\frac{i-1}{N}$ to the sentence as a feature value. For example, when a speech consists of 4 sentences, we assign 0.0, 0.25, 0.5, and 0.75 to the sentences respectively.
- Speaker: In the minutes, there are two types of utterances: (1) utterances that explain activities and (2) utterances that ask questions and request actions. Utterances in the minutes are related to expressions in the summaries. In other words, expressions in utterances are important features for our model. Expressions in the activity explanations differ from the questions and the requests. For example, the activity explanations contain expressions such as "実施 (enforcement)", "策定 (formulation)", and "創設 (establishment)." On the other hand, the questions and the requests contain expression such as "どう (how)", "すべき (should)", "具体的に (concretely)." The difference can be usually distinguished from the Speaker type in the speech (See Table 1.) We define two types of speaker roles: a questioner and a respondent. The type of a speaker in a speech is determined by some rules based on official position information and the political party that the speaker belongs to. We use the Speaker type of a speech as a binary feature.
- Sentence length: It seems that the length of a sentence indicates the information contained in the sentence. We divide the number of characters in the sentence in a speech by the maximum number of characters in all sentences in the speech for a normalization. For example, if a speech consists of sentences containing 100 characters, 50 characters, and 25 characters, we assign 1.0, 0.5, and 0.25 to each sentence, respectively.
- Topic similarity: This feature indicates the similarity between a sentence in a speech and a topic of the speech. Important sentences of a speech may contain relevant elements to the topic of the speech, We calculate similarity scores between each sentence in a speech and a main topic and subtopic of the speech. We assign the scores to each sentence as feature values. We use the cosine similarity between bag-of-words representations of a sentence and a topic as the similarity measure.

3.3   Sentence Extraction

We train an importance prediction model using the features in the previous section with the constructed data explained in Sect. 3.1. The model predicts importance scores of the sentence of an input speech. We extract a set of sentences on the basis of the predicted importance scores. We select sentences in order of predicted importance scores. However, the number of sentences in a reference summary is not given. Therefore, we need to determine the number of sentences in a generated summary. We checked the given corpus. If the limit length is not more than 50 characters, the reference summary tends to consist of one sentence. If the limit length is more than 50 characters, the reference

Table 2. Target expressions and alternative expressions

Target expressions	Alternative expressions
お伺い。	伺う。
を伺う。	は。
と考える。	。
と思う。	。
べきと考えますが、—。	べき。

summary contains multiple sentences. Therefore, we identify the number of sentences on the basis of a limit length L. In our method, we extract $\lfloor \frac{L}{50} \rfloor$ sentences. For example, when the limit length is 100 characters, we extract two sentences.

3.4   Sentence Reduction

When the number of characters contained in extracted sentences is more than the limit length, we apply a sentence reduction process to generate summaries within the limit length. We remove words and chunks for the reduction process in our method.

Algorithm 1. Chunk Deletion

Input: Extracted sentences list $S = [S_0, ..., S_n]$, importance scores of sentences list $I = [I_{S_0}, ..., I_{S_n}]$, and a limit length N_{limit} characters.
Output: *Summary*
1: $listChunks \Leftarrow$ We split S_i into chunks and merge chunks.
   $\{listChunks = [[S_{0_{chunk0}}, ...], ..., [S_{n_{chunk0}}, ...]]\}$
2: **while** $listChunks \neq [[], ..., []]$ **do**
3:    $listIndex \Leftarrow$ Sentence containing the largest number of chunks
      $\{listIndex = [Index_0, ...](0 \leq Index_i \leq n)\}$
4:    **for** $Index \Leftarrow$ (Ascending order of $I_{S_{Index_i}}$) **do**
5:       Removal of $S_{Index_{chunk0}}$
6:       **if** a sum of the number of characters in $listChunks \leq N_{limit}$ **then**
7:          **return** $str(listChunks)$
8:       **end if**
9:    **end for**
10: **end while**
11: **return** $str([])$

Word Deletion. Some expressions at the end of a sentence such as honorific expressions did not appear in reference summaries. Therefore, we remove some expressions on the basis of several rules. The rules are as follow:

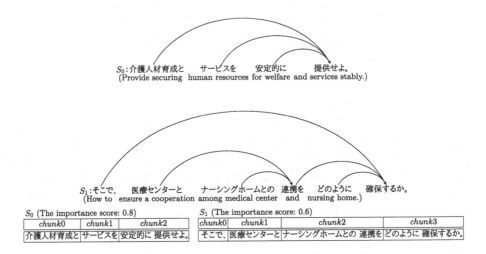

Fig. 3. An example of the sentence reduction process

1. If the last chunk of a sentence contains honorific expressions and contains no negation expressions, we apply a lexical normalization to the chunk. For example, "いたします" is deleted from the original phrase "所見をお伺いいたします。" by this rule.
2. We modify the target expressions in Table 2 to the alternative expressions in Table 2. For example, the chunk "所見をお伺い。" is modified to "所見を伺う。" and "所見を伺う。" is modified to "所見は。" by this rule.

Chunk Deletion. If extracted sentences exceed the limit length even after the processing in the previous section, we delete the chunks to generate a summary within the limit length. In the given corpus in the summarization task at NTCIR-14, an important part of a sentence often appears in the last part of a sentence. Therefore, we suppose that the latter chunks in a sentence are more important than preceding chunks in the sentence. We remove preceding chunks in a sentence. Algorithm 1 shows the algorithm. Figure 3 shows an example of the process.

First, we split extracted sentences into chunks. Next, we merge some chunks to avoid unnatural removal of chunks. If a chunk modifies the chunk immediately after it, we merge the chunks into a single chunk. In Fig. 3, we merge the chunk "安定的に" and the chunk "提供せよ。" into a chunk "安定的に提供せよ。" in the sentence S_0 and merge the chunk "どのように" and the chunk "確保するか。" into a chunk "どのように確保するか。" in the sentence S_1.

We remove preceding chunks to generate a summary within a limit length. When we extract multiple sentences, we remove chunks from sentences containing more chunks preferentially. As shown in the bottom part of Fig. 3, after the chunk merging process, S_0 contains 3 chunks and S_1 contains 4 chunks. Since the number of chunks in S_1 is larger than that in S_0, we remove the chunk "そこで、" in S_1. Then, S_0 and S_1 contain 3 chunks. When the number of chunks is the same, we remove a chunk from a sentence with a lower importance score. When the importance score of S_1 is lower than that of S_0, we remove "医療センターと" in S_1.

4   Experiment

In this section, we describe the experiment to evaluate our method.

4.1   Data

We constructed training data from the NTCIR-14 Summarization Train Dataset. The NTCIR-14 Summarization Train Dataset contains 591 speeches. The speeches contain 9,810 sentences. We applied the process described in Sect. 3.1 to the data. Then, we trained a sentence importance prediction model with the constructed data. For test data, we used the NTCIR-14 Formal Run Dataset. The data contains 146 speeches. The speeches contain 1,759 sentences.

4.2   Baseline Methods

We compared our method with several baseline methods to evaluate our supervised method using constructed training data and our sentence reduction process. In this section, we describe the baseline methods.

For Verification of the Effectiveness of Our Supervised Extractive Approach: We compared our method with methods described in Sect. 2.2. The methods were proposed on the summarization task at NTCIR-14 QA Lab-PoliInfo.

Kimura et al. [3] proposed an abstractive summarization method. We compared our method with the method of Kimura et al. to evaluate the effectiveness of our extractive approach. In addition, we compared our method with unsupervised extractive summarization methods to evaluate our supervised method. The methods are Tang et al. [13], Terazawa et al. [14] and Shinjo et al. [12] described in Sect. 2.2.

For Verification of the Effectiveness of Our Sentence Reduction: We proposed the sentence reduction process in Sect. 3.4. We compared our method with three baseline methods to evaluate the sentence reduction process. First, we compared our method with two supervised methods proposed in the summarization task at NTCIR-14 QA Lab-PoliInfo. The methods used supervised extractive approaches and applied sentence reduction processes as well as our method. We explain the reduction processes.

- Hiai et al. [1]: The method removed words before the first content word and words after the last verbal noun in the sentence.
- Ogawa et al. [10]: The method calculated importance scores of chunks using features such as dependency depth, case information, and frequency in all summaries. The method extracted the chunk with the highest importance score and the other chunks on the path between the chunk with the highest importance score and the last chunk.

However, these two methods often generated ill-formed summaries. Therefore, we compared another strong baseline method with our method. As mentioned in Sect. 2.3, Kiyota et al. [5] reported a dependency-based sentence reduction method. Instead of our sentence reduction method, we applied the Kiyota's method to our extractive approach. In other words, we replaced our sentence redaction process in Fig. 1 to Kiyota's method.

- Our method with Kiyota et al.: First, the method removed conjunctions, adverbs and noun phrases containing time expressions in a sentence. Next, they calculated importance scores of chunks using features such as case components and verbs. The method removed chunks with low importance scores.

Table 3. Results for the evaluation of our supervised extractive approach. Underlines denote the best scores in the results of the methods proposed in the summarization task at NTCIR-14 QA Lab-PoliInfo.

	Recall				F-measure			
	N1	N2	N3	N4	N1	N2	N3	N4
Kimura et al. [3,4]	0.221	0.038	0.013	0.004	0.230	0.038	0.012	0.004
Tang et al. [13]	0.251	0.120	0.079	0.058	0.226	0.107	0.071	0.051
Terazawa et al. [14]	0.400	0.173	0.113	0.076	<u>0.361</u>	<u>0.156</u>	<u>0.102</u>	0.068
Shinjo et al. [12]	0.278	0.060	0.035	0.020	0.240	0.055	0.031	0.018
Our method	**0.475**	**0.229**	**0.155**	**0.112**	**0.394**	**0.187**	**0.123**	**0.088**

Table 4. Results for the evaluation of our sentence reduction. Underlines denote the best scores in the results of the methods proposed in the summarization task at NTCIR-14 QA Lab-PoliInfo.

	Recall				F-measure			
	N1	N2	N3	N4	N1	N2	N3	N4
Hiai et al. [1]	0.390	0.174	0.113	0.078	0.343	0.154	0.101	<u>0.069</u>
Ogawa et al. [10]	<u>0.459</u>	<u>0.200</u>	<u>0.131</u>	<u>0.089</u>	<u>0.361</u>	0.151	0.097	0.064
Our method with Kiyota et al. [5]	0.449	0.195	0.121	0.081	0.370	0.159	0.097	0.065
Our method with our reduction (Full)	**0.475**	**0.229**	**0.155**	**0.112**	**0.394**	**0.187**	**0.123**	**0.088**

Reference Summary
区市と連携し、建築基準法令への適合性の確認により建築物の安全を確保。(By confirming building code compliance, we ensure the safety of the building.)

Extracted Sentence
また、既存建築物については、区市と連携し、定期報告や増築などの機会をとらえ、完了検査の実施状況を把握し、建築基準法令への適合性について確認することにより、建築物の安全確保に努めてまいります。(At the opportunities such as periodic reports and extensions of the buildings, by confirming inspecting situations and by confirming building code compliance, we ensure the safety of the building.)

Kiyota's Reduction
増築などの機会をとらえ、完了検査の実施状況を把握し、確認することにより、建築物の安全確保に努める。(At the opportunities such as extensions of the buildings, by confirming inspecting situations, we ensure the safety of the building.)

Our Reduction
建築基準法令への適合性について確認することにより、建築物の安全確保に努める。(By confirming building code compliance, we ensure the safety of the building.)

Fig. 4. Generated summaries. Underlines denote the expressions contained in the reference summary. Wavy lines denote the expressions not contained in the reference summary.

4.3 Evaluation

The results for the comparison with our method and abstractive and unsupervised methods are shown in Table 3. As the evaluation measure, we used scores in the ROUGE family [7]. Our method outperformed the abstractive method and the unsupervised extraction methods on all criteria. The result shows the effectiveness of our supervised extractive method.

The results for the evaluation of our sentence reduction are shown in Table 4. Our method outperformed the other methods on all criteria. The result shows the effectiveness of our sentence reduction process considering the characteristics of the assembly minutes data.

Additionally, the underlined scores in Tables 3 and 4 denote the best scores of the NTCIR-14 QA Lab-PoliInfo competition. All the scores of our method outperformed the best scores.

5 Discussion

Our method outperformed the method with the sentence reduction based on Kiyota et al. We analyzed the summaries generated with our methods and the method with Kiyota's reduction. Examples of the summaries generated with our methods and the method with Kiyota's reduction are shown in Fig. 4. The summary generated with the method with Kiyota's reduction contains the phrases "増築などの機会 (The opportunities such as extensions of the buildings)" and "完了調査の実施状況を把握 (Confirming inspecting situations)". The phrases appeared in the middle of the extracted sentence. The summary generated with our method contains the phrases contained in the second half of the extracted sentence such as "建築基準法への適合性について確認 (Confirming building code compliance)". As a result, the summary generated with our method was closer to the reference summary than that with the method by Kiyota's reduction. The example shows the effectiveness of our sentence reduction method considering the characteristics of summaries of assembly minutes.

6   Conclusions

Our method extracted important sentences to summarize an assembly member speech in the minutes. Our method used a machine learning approach to predict the importance of sentences. We automatically constructed training data for the sentence importance prediction model construction. We extracted sentences on the basis of the predicted scores and applied the sentence reduction process. In the experiment, we compared our method with baseline methods. The result showed the effectiveness of our method using the supervised importance prediction model trained by the constructed data and the effectiveness of our sentence reduction process. In future work, we need to consider the topic information of a speech to improve our sentence reduction process.

References

1. Hiai, S., Otani, Y., Yamamura, T., Shimada, K.: KitAi-PI: summarization system for NTCIR-14 QA Lab-Poliinfo. In: Proceedings of The 14th NTCIR Conference on Evaluation of Information Access Technologies (2019)
2. Katragadda, R., Pingali, P., Varma, V.: Sentence position revisited: a robust light-weight update summarization baseline algorithm. In: Proceedings of the Third International Workshop on Cross Lingual Information Access: Addressing the Information Need of Multilingual Societies (CLIAWS3), pp. 46–52 (2009)
3. Kimura, T., Tagami, R., Katsuyama, H., Sugimoto, S., Miyamori, H.: KSU systems at the NTCIR-14 QA Lab-Poliinfo task. In: Proceedings of The 14th NTCIR Conference on Evaluation of Information Access Technologies (2019)
4. Kimura, Y., et al.: Overview of the NTCIR-14 QA Lab-Poliinfo task. In: Proceedings of the 14th NTCIR Conference on Evaluation of Information Access Technologies (2019)
5. Kiyota, Y., Kurohashi, S.: Automatic summarization of Japanese sentences and its application to a WWW KWIC index. In: Proceedings of 2001 Symposium on Applications and the Internet (2001)
6. Kubo, T.: MeCab: yet another part-of-speech and morphological analyzer. http://mecab.sourceforge.net/
7. Lin, C.Y.: Rouge: a package for automatic evaluation of summaries. In: Proceedings of the Eighth Workshop on the Annual Meeting of the Association for Computational Linguistics (ACL-04), pp. 74–81 (2004)
8. Mihalcea, R., Tarau, P.: TextRank: bringing order into text. In: Proceedings of the 2004 Conference on Empirical Methods in Natural Language Processing (2004)
9. Mikami, M., Masuyama, S., Nakagawa, S.: A summarization method by reducing redundancy of each sentence for making captions of newscasting. J. Nat. Lang. Process. **6**(6), 65–91 (1999)

10. Ogawa, Y., Satou, M., Komamizu, T., Toyama, K.: nagoy team's summarization system at the NTCIR-14 QA Lab-Poliinfo. In: Proceedings of The 14th NTCIR Conference on Evaluation of Information Access Technologies (2019)
11. Ouyang, Y., Li, W., Lu, Q., Zhang, R.: A study on position information in document summarization. In: Proceedings of the 23rd International Conference on Computational, pp. 919–927 (2010)
12. Shinjo, T., Nishikawa, H., Tokunaga, T.: TTECH at the NTCIR-14 QA Lab-Poliinfo task. In: Proceedings of The 14th NTCIR Conference on Evaluation of Information Access Technologies (2019)
13. Tang, L., Watanabe, K., Yada, S., Kageura, K.: LisLb-Team at the NTCIR-14 QA Lab-Poliinfo task. In: Proceedings of The 14th NTCIR Conference on Evaluation of Information Access Technologies (2019)
14. Terazawa, K., Shirato, D., Akiba, T., Masuyama, S.: AKBL at the NTCIR-14 QA Lab-Poliinfo task. In: Proceedings of The 14th NTCIR Conference on Evaluation of Information Access Technologies (2019)

nagoy Team's Summarization System at the NTCIR-14 QA Lab-PoliInfo

Yasuhiro Ogawa$^{(\boxtimes)}$, Michiaki Satou, Takahiro Komamizu,
and Katsuhiko Toyama

Nagoya University, Nagoya, Japan
yasuhiro@is.nagoya-u.ac.jp

Abstract. The *nagoy* team participated in the NTCIR-14 QA Lab-PoliInfo's summarization subtask. This paper describes our summarization system for assembly member speeches using random forest classifiers. Since we encountered an imbalance in the data, we were unable to achieve good results in this subtask when training on all data. To solve this problem, we developed a new summarization system that applies multiple random forest classifiers training on different-sized data sets step by step. As a result, our system achieved good performance, especially in the evaluation by ROUGE scores. In this paper, we also compare our system with a single random forest classifier using probability.

Keywords: NTCIR-14 · Summarization · Random forest · Progressive ensemble random forest

1 Introduction

According to Alexander Hamilton, the distribution of political information is important to the health of democracies. Although the Japanese government and local governments release political information, it is not enough to utilize for various purposes.

The NTCIR-14's QA Lab-PoliInfo [4] (Question Answering Lab for Political Information) deals with political information and sets forth three tasks: segmentation, summarization, and classification. Our team participated in the summarization task. We previously developed a summarization system [7] for Japanese statutes, which are also political, that is based on random forest classifiers [1] and achieves better results than other summarization systems. Thus, we expected our system to perform equally well for assembly member speeches. However, we were confronted with a data imbalance problem between summarization for statutes and that for assembly member speeches. To overcome this problem, we introduced a new approach that applies multiple random forest classifiers training on different-sized data sets in a step-by-step manner.

This paper describes our summarization system for assembly member speeches and discusses not only the official results, but additional comparison results as well.

© Springer Nature Switzerland AG 2019
M. P. Kato et al. (Eds.): NTCIR 2019, LNCS 11966, pp. 110–121, 2019.
https://doi.org/10.1007/978-3-030-36805-0_9

2 System

Our summarization system consists of two modules: a sentence extraction module using random forest classifiers and a sentence reduction module. In Sect. 2.1, we explain how to construct training data for random forest classifiers and, in Sect. 2.2, show the features we used. We solve the training data imbalance by applying multiple classifiers, as described in Sect. 2.3. In Sect. 2.4, we provide the details of the sentence reduction module.

2.1 Training Data

In this task, a summary of an assembly member's speech is provided as a description of *Togikai dayori*[1]. This summary was not made by sentence extraction methods; that is, a sentence in the summary may not appear in the original speech.

Since our method is based on sentence extraction methods, we need training data that consists of positive and negative sentences, where the "positive" or "negative" sentence means that it is or is not used for making the summary, respectively. Thus, we determined which sentence is used for making a summary as follows.

When we are given a pair consisting of an assembly member's speech and its summary, we find the sentence in the speech that contains the most words in the summary. We consider this sentence to be positive and the others negative. Since this summarization task has a length limit, if the length of the positive sentence is shorter than the length limit, we select the sentence with the second-most summary words. In order to make the training data more correct, we should consider redundancy; that is, we should account for the overlap of the first positive sentence and the second, but we simply choose the second without considering the degree of overlap.

In the formal run, 596 assembly member speeches consisting of 9,979 sentences[2] were provided. Hereafter, we refer to these speeches as the "source documents." Using the above method, we assembled training data that included only 825 (8.3%) positive sentences. This differs considerably from the summarization of Japanese statutes. In our study of statute summarization [7], we used *outlines of Japanese statutes*, which are official summaries of statutes published by the Japanese government, as the gold standard. In this case, the ratio of positive data is over 70% [7]. Because of this difference, we cannot apply the statute summarization methods to assembly member speeches, so we developed another method, described in Sect. 2.3.

2.2 Random Forest Features

In order to train a random forest classifier, we used the following features: sentence position, sentence length, and presence of a word. Here, we selected words

---

[1] https://www.gikai.metro.tokyo.jp/newsletter/.
[2] Two speeches have no sentences.

Table 1. Number of sentences extracted by each classifier

ID	×1	×2	×3	×4	×5	Number of sentences
111	1	0	0	0	0	45
106	9	5	2	0	0	11
19	7	3	3	1	0	8
23	3	2	1	0	0	34
92	5	3	1	1	1	13

that are nouns, occur more than once in the summary, and are not within the top 20 of the number of occurrences in all the source documents.

For the formal run data, we used the presence of 992 words as features.

2.3 Progressive Ensemble Random Forest

Since the above training data includes only 8.3% positive data and is imbalanced, using all of the training data results in poor performance. In fact, when we trained a random forest classifier on all training data, the classifier chose no sentences for 135 of the 146 documents in the test data.

Thus, we utilized an undersampling technique to solve this problem; however, we questioned how much negative data we should use. We prepared the following five random forest classifiers trained on the same positive data with different-sized negative data:

1. classifiers trained equally on positive and negative data,
2. classifiers trained on negative data twice the size of the positive data,
3. classifiers trained on negative data three times the size of the positive data,
4. classifiers trained on negative data four times the size of the positive data,
5. classifiers trained on negative data five times the size of the positive data.

Table 1 shows how many sentences each random forest classifier extracted from the source documents of the test data. In this table, "ID" indicates the identification number of the target documents and "× n" indicates the result of the n-th random forest classifier. ID 111 consists of 45 sentences; the first classifier extracted just one sentence, but other classifiers extracted no sentences. On the other hand, ID 106 consists of 11 sentences and the first classifier extracted 9 sentences, which is too many. In this case, the third classifier, which extracted two sentences, seems better. As can be seen from these results, the most suitable classifier varies from document to document.

Our solution to choosing the classifier is to use all the classifiers step by step, which we call "progressive ensemble."

First, we use the fifth classifier. If that classifier does not extract any sentences, then we use the fourth classifier. If the fourth classifier also extracts no sentences, then we use the third one. We repeat this process until we obtain

Table 2. String replacement as preprocessing

target string	replaced string
でございます。	です。
伺います。	。
ております。	ている。
でおります。	でいる。
てまいります。	ていく。
でまいります。	でいく。
であります。	です。
いたします。	する。
と思います。	。
と思っている。	。

a sentence. Note that we use the next classifier if the length of extracted sentences is ten less than the limit because we find that such extracted sentences are insufficient for summarization. As a result, the length of extracted sentences may exceed the limit.

In addition, the test data in the formal run consists of "single-topic" and "multi-topic" data. We assume that "multi-topic" data needs multiple sentences for summarization, so we choose at least two sentences for "multi-topic" data.

2.4 Sentence Reduction

Since extracted sentences are redundant and sometimes exceed the length limit, we need to reduce them. Our sentence reduction method is a typical one using a Japanese dependency analyzer. We analyze extracted sentences by CaboCha [5] and select the important *bunsetsu* segments (hereafter "segment"). We calculate importance scores using segment features, such as dependency depth, case information, and frequency in all summaries, not used in traditional sentence reduction methods. If a segment contains a noun, its frequency in all summaries of the training data is used as a weight. The weights of other features are adjusted by hand. In particular, we adjusted the weights to improve the ROUGE score between the reduced sentence of a positive sentence in the training data and its corresponding sentence in the summary.

When we reduce an extracted sentence, we first take the last segment. Next we choose the segment with the highest importance score, where we also choose the other segments on the path between the segment with the highest importance score and the last segment to avoid creating ungrammatical sentences. We add the next segments unless the sentence length exceeds the limit.

Although this sentence reduction method always selects the last segment, it is sometimes redundant. Thus, we introduce a replacement process for preprocessing that simply replaces the end of the sentence, as described in Table 2.

活気の vibrant	ある -	社会を society	構築すべきと should build	考えますが think	知事の governor's	所見を伺います。 ~~would like to ask~~	
100	200	300	400	500	500		dependency depth
40	0	20	0	0	40		case information
0	0	13	108	0	61		frequency in all summaries
140	200	333	508	500	601		

Fig. 1. Sentence reduction module

Figure 1 shows an example of sentence reduction. First, we replace the end of the input sentence "活気のある社会を構築すべきと考えますが知事の所見を伺います。(I think we should build a vibrant society and I would like to ask the governor's view.)" according to Table 2, so that the last string "伺います (would like to)" is deleted. Second, we analyze the sentence dependency by CaboCha, so that the sentence is separated into seven segments and the dependency relations are represented by arrows. Third, we calculate importance scores for each segment using dependency depth, case information, and frequency in all summaries. Then we compile segments in order of score. The last segment "所見を (ask)" is chosen first and the sixth segment "知事の (governor's)" is chosen since it has the highest score. Next, we extract the fourth segment "構築すべきと (should build)" with the second-highest score, where we also extract the fifth segment "考えますが (think)" since it is on the path between the fourth segment and the last segment. We repeat this process and choose the third segment. When we select the second segment "ある (-)", the total length of the chosen segments exceeds the limit. Thus, we do not select the second segment and stop the process. As the result, we get the sentence "社会を構築すべきと考えますが知事の所見を (I think we should build a society and I ask the governor's view.)" as a summary.

3   Result of the Formal Run

As an evaluation measure, two scores are used: the scores in the ROUGE [6] family and the scores of the quality questions by the participants [4]. The ROUGE family is ROUGE-N1, -N2, -N3, -N4, -L, -SU4, and -W1.2. The quality questions were assessed by a three-grade evaluation (i.e., A to C) from viewpoints of content, formedness, and total. However, for the content evaluation, an extra grade X was prepared because a summary that does not include gold standard data contents may be acceptable. The quality question score $QQ(v)$ from viewpoint v was calculated using the following expressions:

$$QQ(v) = \frac{\sum_{s \in S} g(s, v)}{|S|}, \tag{1}$$

$$g(s, v) = \begin{cases} 2 & (grade A) \\ 1 & (grade B) \\ 0 & (grade C) \\ a & (grade X) \end{cases}, \tag{2}$$

Table 3. Quality question scores in the formal run (max is 2)

	All-topic				Single-topic				Multi-topic			
	Content		Formed	Total	Content		Formed	Total	Content		Formed	Total
	X = 0	X = 2			X = 0	X = 2			X = 0	X = 2		
KitAi-01	0.856	**1.134**	1.732	**0.912**	**0.953**	1.170	1.660	0.995	0.745	**1.092**	1.815	**0.815**
KitAi-02	0.788	1.035	1.308	0.667	0.849	1.028	1.340	0.722	0.717	1.043	1.272	0.603
TTECH-01	0.290	0.644	1.783	0.402	0.274	0.575	1.755	0.401	0.310	0.723	1.815	0.402
nagoy	**0.886**	1.104	1.619	0.899	**0.953**	**1.179**	1.642	**1.028**	**0.810**	1.016	1.592	0.750
akbl-01	0.722	1.005	1.833	0.826	0.708	1.009	1.844	0.849	0.739	1.000	1.821	0.799
akbl-02	0.707	1.000	1.837	0.793	—	—	—	—	0.707	1.000	1.837	0.793
KSU-01	0.043	0.043	**1.955**	0.048	0.052	0.052	**1.934**	0.057	0.033	0.033	**1.978**	0.038
KSU-02	0.076	0.121	1.745	0.071	0.080	0.156	1.722	0.104	0.071	0.082	1.772	0.033
KSU-03	0.091	0.157	1.715	0.104	0.104	0.179	1.731	0.156	0.076	0.130	1.696	0.043
KSU-04	0.111	0.167	1.419	0.093	0.118	0.193	1.420	0.132	0.103	0.136	1.418	0.049
KSU-05	0.048	0.078	1.692	0.048	0.057	0.085	1.726	0.057	0.038	0.071	1.652	0.038
KSU-06	0.078	0.169	1.535	0.091	0.085	0.151	1.542	0.094	0.071	0.190	1.527	0.087
LisLb-01	0.720	0.942	1.237	0.591	0.722	0.920	1.349	0.684	0.717	0.967	1.109	0.484
TO-01	0.504	0.846	1.763	0.551	0.464	0.794	1.778	0.521	0.550	0.905	1.746	0.586

where S is a set of summaries the participants assessed and a is a constant representing whether acceptable summaries that are different from the gold standard summary are regarded as correct or not. If such summaries are regraded as correct, a is 2; otherwise, a is 0.

Tables 3 and 4 show the official formal run results. Table 3 shows the quality question scores and Table 4 shows the evaluation by ROUGE family scores. Bolded scores reflect the best results among the participants. As shown, our system achieved good performance, especially in the ROUGE scores evaluation. However, the formed score was less than other systems, which indicates that our reduction module created some unnatural sentences.

Figure 2 shows a successful example of our system. In this example, the bolded sentence is extracted and successfully reduced. Figures 3 and 4 show unsuccessful examples of our system. In Fig. 3, although the underlined strings are used in the gold standard summary, our system did not extract them. In the case of Fig. 4, our system successfully extracted the target sentence, but failed to reduce the sentence. Our reduction module deleted the object "めり張り", but left the verb "つけ", which made the reduced sentence unnatural. This is because our module tries to leave as many words as possible within the length limit. To solve this problem, we should delete the verb if we delete its object. In addition, we need to further adjust the weights of the features; for example, in this case we should delete "新年度予算編成作業を進めるべきと考えますが" and leave "めり張りをつけ".

Table 4. ROUGE scores in the formal run (all-topic)

		recall							F-measure						
		N1	N2	N3	N4	L	SU4	W1.2	N1	N2	N3	N4	L	SU4	W1.2
Surface Form	KitAi-01	0.440	0.185	0.121	0.085	0.375	0.217	0.179	0.357	0.147	0.096	0.067	0.299	0.168	0.188
	KitAi-02	0.390	0.174	0.113	0.078	0.320	0.200	0.154	0.343	0.154	0.101	**0.069**	0.281	0.173	0.176
	TTECH-01	0.278	0.060	0.035	0.020	0.216	0.092	0.096	0.240	0.055	0.031	0.018	0.187	0.079	0.111
	nagoy-01	**0.459**	**0.200**	**0.131**	**0.089**	**0.394**	**0.229**	**0.186**	**0.361**	0.151	0.097	0.064	0.305	**0.169**	**0.192**
	akbl-01	0.400	0.173	0.113	0.076	0.345	0.189	0.157	**0.361**	**0.156**	**0.102**	0.068	**0.310**	0.167	0.185
	akbl-02	0.326	0.124	0.080	0.057	0.269	0.147	0.112	0.320	0.119	0.077	0.055	0.262	0.141	0.144
	KSU-01	0.158	0.028	0.009	0.002	0.147	0.043	0.071	0.210	0.039	0.013	0.004	0.196	0.059	0.107
	KSU-02	0.185	0.043	0.021	0.014	0.167	0.063	0.080	0.230	0.056	0.027	0.017	0.209	0.080	0.116
	KSU-03	0.172	0.036	0.008	0.002	0.157	0.050	0.075	0.211	0.043	0.011	0.003	0.192	0.062	0.106
	KSU-04	0.171	0.044	0.013	0.002	0.153	0.055	0.072	0.219	0.056	0.017	0.003	0.195	0.072	0.106
	KSU-05	0.227	0.029	0.010	0.002	0.195	0.064	0.089	0.231	0.029	0.010	0.003	0.196	0.065	0.110
	KSU-06	0.221	0.038	0.013	0.004	0.187	0.065	0.086	0.230	0.038	0.012	0.004	0.192	0.067	0.108
	LisLb-01	0.251	0.120	0.079	0.058	0.211	0.132	0.103	0.226	0.107	0.071	0.051	0.188	0.115	0.118
	TO-01	0.267	0.093	0.061	0.045	0.230	0.117	0.105	0.272	0.086	0.052	0.036	0.233	0.110	0.133
Stem	KitAi-01	0.458	0.199	0.134	0.096	0.389	0.234	0.188	0.373	0.159	0.106	**0.075**	0.311	0.182	0.199
	KitAi-02	0.399	0.179	0.118	0.082	0.326	0.208	0.158	0.351	0.160	0.106	0.074	0.286	0.180	0.181
	TTECH-01	0.289	0.064	0.037	0.022	0.222	0.097	0.099	0.251	0.058	0.033	0.019	0.193	0.084	0.114
	nagoy-01	**0.479**	**0.217**	**0.145**	**0.101**	**0.412**	**0.247**	**0.197**	**0.377**	0.165	0.108	0.074	0.319	**0.184**	**0.205**
	akbl-01	0.415	0.184	0.122	0.083	0.357	0.203	0.164	0.375	**0.165**	**0.110**	0.074	**0.322**	0.179	0.195
	akbl-02	0.339	0.135	0.089	0.064	0.279	0.158	0.119	0.333	0.129	0.085	0.063	0.272	0.152	0.153
	KSU-01	0.161	0.028	0.010	0.002	0.148	0.044	0.071	0.214	0.040	0.013	0.004	0.197	0.061	0.108
	KSU-02	0.187	0.044	0.021	0.014	0.170	0.064	0.081	0.233	0.057	0.027	0.017	0.212	0.082	0.117
	KSU-03	0.175	0.036	0.008	0.002	0.159	0.052	0.075	0.217	0.044	0.011	0.003	0.196	0.065	0.108
	KSU-04	0.174	0.045	0.014	0.002	0.155	0.056	0.073	0.222	0.058	0.018	0.003	0.197	0.073	0.107
	KSU-05	0.230	0.029	0.010	0.002	0.199	0.066	0.090	0.236	0.030	0.010	0.003	0.201	0.067	0.112
	KSU-06	0.226	0.040	0.013	0.004	0.189	0.066	0.087	0.235	0.039	0.012	0.004	0.195	0.069	0.109
	LisLb-01	0.261	0.125	0.084	0.061	0.218	0.139	0.106	0.235	0.112	0.075	0.055	0.195	0.121	0.122
	TO-01	0.273	0.097	0.065	0.048	0.233	0.121	0.107	0.277	0.089	0.056	0.039	0.236	0.114	0.136
Content Word	KitAi-01	0.285	0.145	0.090	0.050	0.278	0.154	0.180	0.224	0.115	**0.071**	0.042	0.217	0.107	0.170
	KitAi-02	0.254	0.126	0.083	**0.053**	0.247	0.131	0.156	0.214	0.109	0.069	**0.046**	0.208	0.106	0.159
	TTECH-01	0.088	0.028	0.015	0.007	0.082	0.033	0.050	0.076	0.024	0.012	0.006	0.071	0.027	0.054
	nagoy-01	**0.326**	**0.164**	**0.094**	0.046	**0.315**	**0.168**	**0.201**	**0.249**	**0.123**	0.067	0.036	**0.239**	**0.110**	**0.187**
	akbl-01	0.256	0.113	0.065	0.034	0.247	0.124	0.148	0.224	0.098	0.056	0.031	0.216	0.100	0.158
	akbl-02	0.200	0.094	0.051	0.032	0.189	0.095	0.109	0.188	0.089	0.049	0.031	0.178	0.087	0.127
	KSU-01	0.048	0.001	0.000	0.000	0.047	0.007	0.032	0.059	0.001	0.000	0.000	0.058	0.009	0.043
	KSU-02	0.069	0.014	0.000	0.000	0.067	0.019	0.043	0.083	0.015	0.000	0.000	0.081	0.022	0.059
	KSU-03	0.041	0.002	0.000	0.000	0.041	0.007	0.027	0.050	0.002	0.000	0.000	0.050	0.008	0.036
	KSU-04	0.050	0.002	0.000	0.000	0.048	0.008	0.031	0.064	0.003	0.000	0.000	0.061	0.011	0.044
	KSU-05	0.067	0.002	0.000	0.000	0.062	0.013	0.041	0.063	0.003	0.000	0.000	0.057	0.011	0.043
	KSU-06	0.053	0.003	0.000	0.000	0.051	0.008	0.034	0.051	0.003	0.000	0.000	0.049	0.009	0.037
	LisLb-01	0.171	0.083	0.044	0.026	0.160	0.088	0.106	0.140	0.068	0.036	0.023	0.130	0.065	0.102
	TO-01	0.116	0.055	0.035	0.012	0.111	0.056	0.070	0.106	0.042	0.023	0.011	0.101	0.042	0.076

Source Document	次に、新しい公共について伺います。 新しい公共という考え方は、私たちが国家戦略の柱として、地域主権改革とともに、これからのあるべき社会像として掲げたものです。 日本では、古くから連、結、講、座、あるいは若者組などの住民組織や市井の寺子屋、隠居という名のボランティア的な活動などが活力ある市民社会を担っていました。 新しい公共の考え方は、以前あったこのような社会を現在にふさわしい形で再構築することを目指すものです。 東日本大震災の被災地では、数々のボランティア活動が行われています。 強制ではなくみずからの意思で支援活動をされていた多くの方々の姿は感動的であり、改めて人々のつながりと助け合いの大切さを感じさせられました。 石原都知事は、都の防災対応指針において、自助、共助の徹底について述べられています。 行政依存ではなく、一人一人自立した個が、地域、社会を主体的に働きかけていく協働は、災害時には不可欠なものです。 そこで伺います。 **東京都においては、このような新しい公共型社会の実現を目指し、支え合いと活気のある社会を構築していくべきと考えますが、知事の所見を伺います。**
System Output	公共型社会の実現を目指し、支え合いと活気のある社会を構築していくべきと考えますが、知事の所見を。
Gold Standard	支え合いと活気のある社会を構築すべき。知事の所見を。

Fig. 2. Example of successful summarization

Source Document	次に、雇用創出と学生の就職に向けた取り組みについてであります。 都はこれまでも、厳しい雇用情勢に対応し、失業された方々に対する雇用創出を図るため、区市町村とも連携して、昨年度を大きく上回る約一万八千人の規模で、緊急雇用創出事業に取り組んでまいりました。 しかしながら、雇用情勢は依然として厳しく、ご指摘のとおり、失業された方々に対し、速やかに新たな雇用創出を図ることが重要でございます。 このため、都は区市町村や民間事業者と連携し、年末から年度末に向けた新たな事業を 十億円規模で 追加実施 することによりまして、さらなる雇用創出を図ってまいります。 また、厳しい状況に置かれた学生の就職を後押しするため、都は都内 経済団体に対し採用拡大の要請を、本年七月に続きまして、年内にも再度実施いたします。 さらに、都内中小企業と新規学卒者等とのマッチングの場である合同就職面接会を、既に七月と十一月に**開催し、合わせて三千六百十一人が参加いただいたところでありますが、来年二月にも開催を予定しております。** 今後とも、現下の厳しい雇用情勢に迅速かつ的確に対応してまいります。
System Output	開催し、合わせて三千六百十一人が参加いただいたところでありますが、来年二月にも開催を予定している。
Gold Standard	年末から年度末に新事業を追加実施。経済団体に学生の採用拡大を要請。

Fig. 3. Example of unsuccessful extraction

4 Discussion

4.1 Experimental Discussion

In the traditional sentence extraction summarization method, all sentences in the source documents are scored and are selected in the order of their scores until the

Source Document	我が国の経済にあっては、欧州の債務危機や歴史的な円高などが、回復の兆しが見えた景気に冷や水を浴びせています。企業収益の動向は不透明さを増しており、今後の都税収入への影響は避けられません。こうした中、都には、少子高齢化や中小企業対策など、山積する課題に対して効果的な手だてを講じ、現下の閉塞感を打ち破り、東京に活力を呼び戻していくことが求められています。とりわけ、震災への対応は喫緊の課題です。我が党が立ち上げた東日本大震災復旧・復興対策推進本部で議論を重ね、先月、防災力強化に向けての提言を行いました。提言内容も含め、高度防災都市の実現に向けた取り組みを加速する上では、法人事業税の暫定措置の撤廃は不可欠であり、約束どおり撤廃するよう国に強く求めるものであります。この間、国が公共事業を見識ある考えもなく削減し続けたのとは対照的に、都は七年連続で投資的経費を伸ばしてきました。都税収の回復が当面期待できない今だからこそ、中小企業の受注機会をふやすなど、景気を刺激し、防災力強化にも資する投資的経費に財源を振り向けることが重要であります。**これまで以上にめり張りをつけ、都民に安心と希望をもたらす予算とするべく、新年度予算編成作業を進めるべきと考えますが、所見を伺います。**
System Output	つけ、都民に安心と希望をもたらす予算とするべく、新年度予算編成作業を進めるべきと考えますが、所見を。
Gold Standard	メリハリをつけて、都民に安心と希望をもたらす予算にするべき。所見は。

Fig. 4. Example of unsuccessful sentence reduction

given length limit is reached. This technique often considers redundancy such as Maximal Marginal Relevance [2].

Although classifiers such as random forest or support vector machine (SVM) basically return a binary or multivalued output, many classifier implementations can return a continuous output. Thus, we can use such output as a score for summarization. For example, a summarization method using an SVM classifier uses the distance from the hyperplane [3].

We used scikit-learn's random forest classifier, which can return the probability of each sample. Thus, we compared the method using the probability with our proposed method. We used the same classifiers in the formal run and added one more classifier trained on all data without undersampling. We modified the classifiers to output the probability and chose the sentences in the order of their probabilities. We chose extra sentences if the length of the chosen sentences was ten less than the limit, as with the proposed method.

We conducted the extraction tests with these classifiers and calculated precision, recall, and F-measure, as shown in Table 5. In these tests, we considered the positive sentences gathered by the method described in Sect. 2.1 to be the gold standard. In Table 5, "closed" indicates the closed test conducted on the training data in the formal run. Similarly, "open" indicates the open test conducted on the test data in the formal run, where the test data consists of "single-topic" and "multi-topic" data.

Table 5. Extraction results

			Proposed	×1	×2	×3	×4	×5	All
Precision	closed		0.963	0.860	0.967	0.973	0.983	0.987	**1.00**
	open	all	0.446	0.465	0.471	0.520	**0.526**	0.511	0.523
		single	0.481	0.482	0.464	0.560	**0.588**	0.553	0.571
		multi	0.417	0.450	0.477	**0.483**	0.466	0.473	0.477
Recall	closed		**0.967**	0.785	0.893	0.875	0.886	0.886	0.896
	open	all	**0.523**	0.437	0.406	0.452	0.462	0.457	0.457
		single	**0.526**	0.432	0.411	0.495	0.526	0.495	0.505
		multi	**0.520**	0.441	0.402	0.412	0.402	0.422	0.412
F-measure	closed		**0.965**	0.821	0.929	0.921	0.932	0.933	0.945
	open	all	0.481	0.450	0.436	0.484	**0.492**	0.483	0.488
		single	0.503	0.456	0.436	0.525	**0.556**	0.522	0.536
		multi	**0.463**	0.446	0.436	0.444	0.432	0.446	0.442

Table 5 shows that the first and second classifier scored low because of the low amount of their training data. With precision, although the classifier using all the training data scored the highest in the closed test, we consider this overfitting. The fourth classifier achieved the highest precision score in the open test. With recall, our proposed method scored the highest in all tests. Since the recall score is more important in summarization, our method is more suitable and thus our system achieved good performance in the formal run.

Another advantage of our method is that it does not need to tune the balance between positive and negative data. Although Table 5 shows that the fourth classifier achieved the highest precision and F-measure scores, this does not always hold. The third classifier might achieve higher performance when we use other training data. However, our method does not need to consider which ratio of positive and negative data is best, and uses multiple training data with different ratios.

Source Document	最後に、医療技術分野の産業振興についてでありますが、都は、今後の成長が**期待できる医療の分野を重点的に育成する**との方針のもと、中小企業の **製品開発**や事業化を既に **支援して** おります。
System Output	期待できる医療の分野を重点的に育成するとの方針のもと、中小企業の製品開発や事業化を既に支援している。
Gold Standard	製品開発等を支援している。

Fig. 5. Example of too short summary in the gold standard (1)

Source Document	都内四十七カ所の病院などが**実施している無料低額診療事業の役割は重要です**。 実施病院で話を伺いましたが、ぐあいが悪くても病院に通うお金がなく、無料低額診療でやっと医療に結びついた方がたくさんいる、最近は若い人もふえているとのことでした。 地域におけるセーフティーネットの役割を果たしています。 無料低額診療事業の重要性をどう考えていますか。 実施医療機関はふえているのではありませんか。 さらにふやす必要がある と思います。 都立病院や**公社病院でも実施できると思いますが、いかがですか**。
System Output	実施している無料低額診療事業の役割は重要です。公社病院でも実施できると思いますが、いかがですか。
Gold Standard	更に増やす必要がある。

Fig. 6. Example of too short summary in the gold standard (2)

Source Document	こうした社会的事業は、地域雇用の創出にもつながり、地域を元気にする新たな事業として期待されます。 しかし、立ち上げ時には経営に関する知識や事業分野の専門性にも乏しいことなどから、十分な事業経費や人件費を得ることは容易ではありません。 事業として継続していけるよう、さまざまな角度からの支援や社会的仕組みが墨田で始まっています。 私の地元多摩地域でも、市民が主体となった事業が芽吹いており、多摩地域への開設を要望するものです。 ソーシャルビジネスを展開する団体に対し、活動拠点の確保や**事業運営に必要なスキルを磨く拠点や機能を提供するような努力を積極的に行うべきと考えますが、所見を伺います**。
System Output	事業運営に必要なスキルを磨く拠点や機能を提供するような努力を積極的に行うべきと考えますが、所見を。
Gold Standard	様々な角度からの支援を。

Fig. 7. Example of unsuitable summary in the gold standard

4.2   Quality of the Gold Standard

In this task, we consider that a description of *Togikai dayori* is a gold standard summary for an assembly member's speech, but some are too short and not suitable, as shown in Figs. 5 and 6. In Fig. 5, the gold standard summary "製品開発等を支援している (We are supporting product development, etc.)" omits what products to develop and, in Fig. 6, "更に増やす必要がある (We need to increase more)" omits what should be increased. Although both summaries are abstract, our system outputs more concrete summaries. Our summary includes the gold standard in the former, but not the latter, case. Although we consider our summary in the latter case to be suitable, it reduces the ROUGE score. In the quality question scores, the latter case was evaluated as grade X, as described in Sect. 3.

A more extreme case is shown in Fig. 7. In this example, the gold standard summary "様々な角度からの支援を (The support from various angles)"

is considered as a request. However, this phrase is extracted from the sentence "事業として継続していけるよう、さまざまな角度からの支援や社会的仕組みが墨田で 始まっています。 (The support from various angles and social mechanisms has begun in Sumida so that we can continue with them as a business)," where "the support from various angles" is not a request but a fact that has begun. Our system extracted a different sentence and was evaluated as grade X; that is, a human evaluator judged it is a suitable summary.

From this, we can conclude that we should pay attention to summaries in the gold standard data that are too short as they may adversely affect not only tests, but also training.

5   Conclusions

This paper described our summarization system at the NTCIR-14 QA Lab-PoliInfo. We proposed a progressive ensemble random forest method that applies multiple random forest classifiers training on different-sized data sets step by step in order to deal with imbalanced data. Although we achieved good performance, especially in the evaluation by ROUGE scores, our sentence reduction module sometimes caused our system to create unnatural sentences.

Thus, our future work is to improve the sentence reduction module. We would also like to investigate the relationship between our progressive ensemble random forest classifiers and the probability they estimated.

Acknowledgments. This work was partly supported by JSPS KAKENHI Grant Number 17K00460.

References

1. Breiman, L.: Random forests. Mach. Learn. **45**(1), 5–32 (2001)
2. Carbonell, J., Goldstein, J.: The use of MMR, diversity-based reranking for reordering documents and producing summaries. In: Proceedings of ACM-SIGIR 1998, pp. 335–336 (1998)
3. Hirao, T., Isozaki, H., Maeda, E., Matsumoto, Y.: Extracting important sentences with support vector machines. In: Proceedings of the 19th International Conference on Computational Linguistics, vol. 1, pp. 1–7 (2002)
4. Kimura, Y., et al.: Overview of the NTCIR-14 QA Lab-PoliInfo task. In: Proceedings of the 14th NTCIR Conference (2019)
5. Kudo, T., Matsumoto, Y.: Japanese dependency analysis using cascaded chunking. In: Proceedings of the 6th Conference on Natural Language Learning 2002 (COLING 2002 Post-Conference Workshops), CoNLL 2002, pp. 63–69 (2002)
6. Lin, C.-Y.: ROUGE: a package for automatic evaluation of summaries. In: Proceedings of the ACL 2004 Workshop, vol. 8, pp. 74–81 (2004)
7. Ogawa, Y., Satou, M., Komamizu, T., Toyama, K.: Extracting important sentences with random forest for statute summarization. In: Proceedings of the Annual Conference of JSAI 2019 (2019). (In Japanese)

Final Report of the NTCIR-14 QA Lab-PoliInfo Task

Yasutomo Kimura[1,2], Hideyuki Shibuki[3,4(✉)], Hokuto Ototake[5], Yuzu Uchida[6], Keiichi Takamaru[7], Kotaro Sakamoto[3,4], Madoka Ishioroshi[3], Teruko Mitamura[8], Noriko Kando[3,9], Tatsunori Mori[4], Harumichi Yuasa[10], Satoshi Sekine[2], and Kentaro Inui[2,11]

[1] Otaru University of Commerce, Otaru, Japan
[2] RIKEN, Wako, Japan
[3] National Institute of Informatics, Tokyo, Japan
shib@nii.ac.jp
[4] Yokohama National University, Yokohama, Japan
[5] Fukuoka University, Fukuoka, Japan
[6] Hokkai-Gakuen University, Sapporo, Japan
[7] Utsunomiya Kyowa University, Utsunomiya, Japan
[8] Carnegie Mellon University, Pittsburgh, USA
[9] SOKENDAI, Hayama, Japan
[10] Institute of Information Security, Yokohama, Japan
[11] Tohoku University, Sendai, Japan

Abstract. The NTCIR-14 QA Lab-PoliInfo aims to achieve real-world complex question-answering (QA) technologies using Japanese political information, such as local assembly minutes and newsletters. QA Lab-PoliInfo has three tasks, namely, segmentation, summarization and classification. We describe the dataset used, formal run results, and comparison between human marks and automatic evaluation scores.

Keywords: NTCIR-14 · QA Lab · PoliInfo · Question answering · Political information · Local assembly minutes · Segmentation · Summarization · Classification

1 Introduction

The Question-Answering Lab for Political Information (QA Lab-PoliInfo) at NTCIR 14 aims to achieve complex real-world question-answering (QA) technologies to show summaries of the opinions of assembly members and the reasons and conditions for such opinions from Japanese regional assembly minutes.

We reaffirm the importance of fact-checking because of the negative impact of fake news in recent years. The International Fact-Checking Network of the Poynter Institute established in 2017 that April 2 would be considered International Fact-Checking Day. In addition, fact-checking is difficult for general Web search engines to deal with because of the filter bubble developed by Pariser [1], which

© Springer Nature Switzerland AG 2019
M. P. Kato et al. (Eds.): NTCIR 2019, LNCS 11966, pp. 122–135, 2019.
https://doi.org/10.1007/978-3-030-36805-0_10

	Fake News Challenge Stage 1	CLEF-2018 Fact Checking Lab	NTCIR QA Lab-PoliInfo
Dataset	News articles	Political debate	Assembly minutes and newsletter
Task	Stance Detection Classifying the stance of the body using both a headline and a body text. Output is as follows: 1. Agree 2. Disagree 3. Discussed 4. Unrelated	Task1: Check-worthiness Prediction which claim in a political debate should be prioritized for fact-checking. Task2: Factuality Checking the factuality of the identified worth-checking claims.	Task1: Segmentation Extracting of the range of primary information Task2: Summarization Summarizing of local assembly member's and governor's utterance Task3: Classification Classifying an utterance which includes fact-checkable statement and opinion for a political topic.
Number of training data	2,586 articles	1,400 sentences x 3 files	Segmentation : 298 set Summarization : 596 set Classification : 14 topic (includes 10,291 sentences)
Language	English	English and Arabic	Japanese

Fig. 1. Comparison with related shared tasks

prevents users from accessing information that disagrees with their viewpoints. For fact-checking, we should confirm data with primary sources, such as assembly minutes. The description of the Japanese assembly minutes is a transcript of a speech, which is very long; therefore, understanding the content, including the opinions of the assembly members, at a glance is difficult. New information-access technologies to support user understanding are expected to be developed, which would protect us from fake news.

We provide Japanese Regional Assembly Minutes Corpus as the training and test data, and we investigate the appropriate evaluation metrics and methodologies for structured data as a joint effort of participants.

QA using the Japanese regional assembly minutes has the following challenges to consider:

(1) a comprehensible summary of a topic;
(2) the beliefs and attitudes of assembly members;
(3) the mental spaces of other assembly members;
(4) the context, including reasons;
(5) several topics in a speech; and
(6) colloquial Japanese, including dialect and slang.

In addition to QA technologies, this task will contribute to the development of semantic representation, context understanding, information credibility, automated summarization, and dialog systems.

2 Related Work

The Fake News Challenge[1] and CLEF-2018 Fact Checking Lab[2] are shared tasks dealing with political information. The stance detection task is conducted in the Fake News Challenge to estimate the relative perspective (or stance) of two pieces of text relative to a topic, claim, or issue. CLEF-2018 Fact Checking Lab conducted the check-worthiness and factuality tasks. Figure 1 shows a comparison of the related shared tasks.

---

[1] http://www.fakenewschallenge.org/.
[2] http://alt.qcri.org/clef2018-factcheck/.

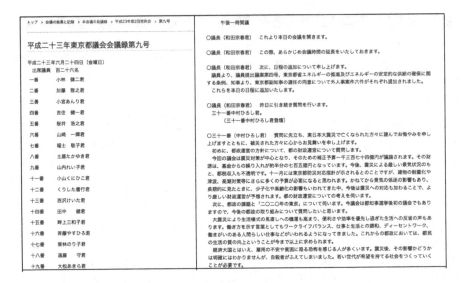

Fig. 2. Example of plenary minutes of the Tokyo Metropolitan Assembly

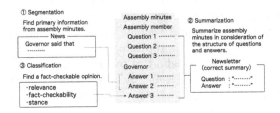

Fig. 3. Relation of the three tasks

3   Japanese Regional Assembly Minutes Corpus

Kimura et al. [4] constructed the Japanese Regional Assembly Minutes Corpus that collects the minutes of plenary assemblies in 47 prefectures of Japan from April 2011 to March 2015. Figure 2 shows an example of the minutes of the Tokyo Metropolitan Assembly. The Japanese minutes resemble a transcript. In the question-and-answer session, a member of assembly asks several questions at a time, and a prefectural governor or a superintendent answers the questions under his/her charge at a time. A speech is very long for the readers to understand the contents at a glance; therefore, information-access technologies such as QA and automated summarization will aid in understanding. For the QA Lab-PoliInfo task, we distributed a subset of the corpus, which is narrowed down to the Tokyo Metropolitan Assembly.

4   Task Description

We designed the segmentation, summarization and classification tasks. We put the tasks at the elemental technologies of information credibility or fact-checking

Table 1. Data fields used in the Segmentation task

Field name	Explanation	Dry run	Formal run
ID	Identification code	○	○
Prefecture	Prefecture name	○	○
date	According to the Japanese calendar	○	○
Meeting	According to *Togikai dayori*	○	○
MainTopic	According to *Togikai dayori*	○	○
SubTopic	According to *Togikai dayori*	○	○
Speaker	Name of member of assembly	○	-
Summary	Description in *Togikai dayori*	○	-
QuestionSpeaker	Name of member of assembly	-	○
QuestionSummary	Description in *Togikai dayori*	-	○
AnswerSpeaker	Name of member of assembly	-	○
AnswerSummary	Description in *Togikai dayori*	-	○
StartingLine	**Answer section**	○	-
EndingLine	**Answer section**	○	-
QuestionStartingLine	**Answer section**	-	○
QuestionEndingLine	**Answer section**	-	○
AnswerStartingLine	**Answer section**	-	○
AnswerEndingLine	**Answer section**	-	○

for political information systems. Figure 3 shows the relation of the tasks. The segmentation task aims to find primary information corresponding to the given secondary information. The summarization task aims to generate brief texts considering argument structures, such as questions and answers. The classification task aims to find pros and cons of a political topic and present their fact-checkable reasons. We preliminarily conducted the tasks in a dry run. We discussed the results with participants in two round-table meetings and refined the tasks for the formal run. The task was conducted in Japanese only because we could not prepare minutes in other languages. This paper describes the task description at the formal run (Table 1).

4.1 Segmentation Task

For the segmentation task, the minutes of the Tokyo Metropolitan Assembly from April 2011 to March 2015 and a summary of a speech of a member of assembly described in *Togikai-dayori*[3], a public relations paper of the Tokyo Metropolitan Assembly are provided. The participants find the corresponding original speech from the minutes and answer the positions of the first and last sentences of the found speech. As an evaluation measure, we used the recall R_{seg},

---

[3] https://www.gikai.metro.tokyo.jp/newsletter/ (in Japanese).

Table 2. Data fields used in the summarization task

Field name	Explanation	Dry run	Formal run
ID	Identification code	○	○
Prefecture	Prefecture name	○	○
date	According to the Japanese calendar	○	○
Meeting	According to *Togikai-dayori*	○	○
Speaker	Name of member of assembly	○	○
StartingLine	The number of first sentence	○	○
EndingLine	The number of last sentence	○	○
MainTopic	According to *Togikai-dayori*	○	○
SubTopic	According to *Togikai-dayori*	○	○
Summary	**Answer section**	○	○
Length	Limit length	○	○
Source	Speech of member of assembly	○	○

precision P_{seg}, and F-measure F_{seg} of concordance of the first and last sentences to the gold standard data. They were calculated using the following expressions:

$$R_{seg} = \frac{N_{cp}}{N_{gsp}}, \tag{1}$$

$$P_{seg} = \frac{N_{cp}}{N_{sp}}, \tag{2}$$

$$F_{seg} = \frac{2R_{seg}P_{seg}}{R_{seg} + P_{seg}}, \tag{3}$$

where N_{cp} is the number of the first and last sentences of which the position is in agreement with the gold-standard position, N_{gsp} is the number of the gold-standard positions, and N_{sp} is the number of sentence positions the participants submitted. In the round-table meetings after the dry run, the participants reported that other sentences had almost the same meaning as the gold-standard statement. We distinguished them by refining the input of a single speech to a pair of question and answer speeches for the formal run (Table 2).

Input: The minutes and a pair of summaries of a question and the answer of a member of assembly
Output: The first and the last sentences of the original speech corresponding to each summary
Evaluation: Recall, precision, and F-measure of the concordance rate of the first and last sentences.

4.2   Summarization Task

For the summarization task, an assembly member's speech and the length limit of the summary were given. The participants generated a summary correspond-

Table 3. Data fields used in Classification task

Field name	Explanation	Dry run	Formal run
ID	Identification code	○	○
Topic	Political topic	○	○
Utterance	A sentence in the minutes	○	○
Relevance	**Answer section**	-	○
Fact-checkability	**Answer section**	-	○
Stance	**Answer section**	-	○
Class	**Answer section**	○	○

ing to the speech within the limit length. As an evaluation measure, we used the scores in the ROUGE [5] family and the scores of the quality questions by the participants. The ROUGE family denotes ROUGE-N1, -N2, -N3, -N4, -L, -SU4, and -W1.2. The quality questions were assessed by a three-grade evaluation (i.e., A to C) in terms of content, formedness, and total. However, for the content evaluation, we prepared an extra grade X because a summary that does not include the contents of gold-standard data may be acceptable. The quality-question score $QQ(v)$ from viewpoint v was calculated using the following expressions:

$$QQ(v) = \frac{\sum_{s \in S} g(s, v)}{|S|}, \qquad (4)$$

$$g(s, v) = \begin{cases} 2 \ (grade A), \\ 1 \ (grade B), \\ 0 \ (grade C), \\ a \ (grade X), \end{cases} \qquad (5)$$

where S is a set of summaries the participants assessed, and a is a constant representing whether acceptable summaries that are different from the gold-standard summary are regarded as correct or not. If such summaries are considered correct, a is 2; otherwise, a is 0 (Table 3).

Input: An assembly member's speech in the minutes and the length limit of the summary
Output: A summary corresponding to the speech
Evaluation: ROUGE scores and participants assessment in terms of content, formedness and total.

4.3 Classification Task

For the classification task, a political topic, such as "The Tsukiji Market should move to Toyosu," and a sentence in the minutes are given. The participants classify the sentence into the following three classes: support with fact-checkable reasons (S), against with fact-checkable reasons (A), and other (O). As evaluation

Table 4. Active participating teams

Team ID	Organization
FU01*	Fukuoka University
FU02*	Fukuoka University
KitAi	Kyushu Institute of Technology
TTECH	Tokyo Institute of Technology
nami	Hitachi, Ltd.
nagoy	Nagoya University
akbl	Toyohashi University of Technology
ibrk	Ibaraki University
RICT	Ricoh Company, Ltd.
STARS	Hokkaido University
tmcit	Tokyo Metropolitan College of Industrial Technology
KSU	Kyoto Sangyo University
CUTKB	University of Tsukuba
LisLb	University of Tokyo
TO*	Task Organizers

*Task organizer(s) are in the team

measures, we used the accuracy of all classes A. Then, the recall $R_{cla}(c)$, precision $P_{cla}(c)$ and F-measure $F_{cla}(c)$ were used for each class c.

$$A = \frac{N_{cc}}{N_{ca}}, \tag{6}$$

$$R_{cla}(c) = \frac{N_{cc}(c)}{N_{gsc}(c)}, \tag{7}$$

$$P_{cla}(c) = \frac{N_{cc}(c)}{N_{sc}(c)}, \tag{8}$$

$$F_{cla}(c) = \frac{2R_{cla}(c)P_{cla}(c)}{R_{cla}(c) + P_{cla}(c)}, \tag{9}$$

where N_{acc} is the number of sentences for which the classified class is in agreement with the gold-standard class; N_{asc} is the number of all sentences, $N_{cc}(c)$ is the number of sentences, of which the gold-standard class is c, that is classified into c, $N_{gsc}(c)$ is the number of sentences of which the gold-standard class is c, and $N_{sc}(c)$ is the number of sentences classified into c. In the round-table meetings after Dry run, we discussed basic factors of classification with participants, and agreed that the factors were relevance, fact-checkability and stance. The relevance means whether or not a given sentence refers to a given topic. The fact-checkability means whether or not the sentence contains fact-checkable reasons. The stance means whether or not a speaker of the sentence agrees on the topic. However, we prepared the third stance, other (O), if a speaker stands

Table 5. Result of segmentation task in formal run

	R		P		F
nami-01	0.814	(1,433/1,761)	0.940	(1,433/1525)	0.872
nami-02	0.864	(1,521/1,761)	0.851	(1,521/1,788)	0.857
nami-03	0.984	(1,733/1,761)	0.499	(1,733/3,475)	0.662
nami-04	0.639	(1,125/1,761)	0.805	(1,125/1,398)	0.712
nami-05	0.553	(973/1,761)	0.931	(973/1,045)	0.694
nami-06	0.655	(1,153/1,761)	0.657	(1,153/1,754)	0.656
nami-07	0.797	(1,404/1,761)	0.933	(1,404/1,505)	0.860
nami-08	0.831	(1,464/1,761)	0.932	(1,464/1,570)	0.879
nami-09	0.875	(1,541/1,761)	0.843	(1,541/1,827)	0.859
nami-10	0.993	(1,749/1,761)	0.464	(1,749/3,769)	0.632
nami-11	1.000	(1,761/1,761)	0.112	(1,761/15,765)	0.201
akbl-01	0.768	(1,352/1,761)	0.538	(1,352/2,515)	0.633
akbl-02	0.847	(1,492/1,761)	0.455	(1,492/3,282)	0.592
akbl-03	0.656	(1,155/1,761)	0.519	(1,155/2,227)	0.580
RICT-01	0.882	(1,554/1,761)	0.909	(1,554/1,709)	0.895
RICT-02	0.856	(1,507/1,761)	0.889	(1,507/1,695)	0.872
RICT-03	0.853	(1,503/1,761)	0.780	(1,503/1,926)	0.815
RICT-04	0.780	(1,374/1,761)	0.746	(1,374/1,842)	0.763
RICT-05	0.936	(1,648/1,761)	0.712	(1,648/2,314)	0.809
KSU-01	0.779	(1,372/1,761)	0.243	(1,372/5,643)	0.370
KSU-02	0.759	(1,337/1,761)	0.268	(1,337/4,998)	0.396
KSU-03	0.820	(1,444/1,761)	0.661	(1,444/2,185)	0.732
KSU-04	0.797	(1,403/1,761)	0.922	(1,403/1,521)	0.855
TO-01	0.354	(623/1,761)	0.898	(623/694)	0.508

neutral or has no relation to the topic. For the formal run, we refined the output to the factors besides class.

Input: A political topic and a sentence in the minutes
Output: The relevance (existence or absence), fact-checkability (existence or absence), stance (agree, disagree, or other), and class (support with fact-checkable reasons, against with fact-checkable reasons, or other)
Evaluation: The accuracy of all classes, and the recall, precision, and F-measure of each class.

5 Result

Fifteen teams described in Table 4 participated, and 83 runs were submitted in total. For the segmentation task, 24 runs from 5 teams were submitted. For the

Table 6. Quality question scores in formal run (max is 2)

	All-topic				Single-topic				Multi-topic			
	Content		Formed	Total	Content		Formed	Total	Content		Formed	Total
	$X=0$	$X=2$			$X=0$	$X=2$			$X=0$	$X=2$		
KitAi-01	0.856	_1.134_	1.732	_0.912_	_0.953_	1.170	1.660	0.995	0.745	_1.092_	1.815	_0.815_
KitAi-02	0.788	1.035	1.308	0.667	0.849	1.028	1.340	0.722	0.717	1.043	1.272	0.603
TTECH-01	0.290	0.644	1.783	0.402	0.274	0.575	1.755	0.401	0.310	0.723	1.815	0.402
nagoy-01	_0.886_	1.104	1.619	0.899	_0.953_	_1.179_	1.642	_1.028_	_0.810_	1.016	1.592	0.750
akbl-01	0.722	1.005	1.833	0.826	0.708	1.009	1.844	0.849	0.739	1.000	1.821	0.799
akbl-02*	0.707	1.000	1.837	0.793	—	—	—	—	0.707	1.000	1.837	0.793
KSU-01	0.043	0.043	_1.955_	0.048	0.052	0.052	_1.934_	0.057	0.033	0.033	_1.978_	0.038
KSU-02	0.076	0.121	1.745	0.071	0.080	0.156	1.722	0.104	0.071	0.082	1.772	0.033
KSU-03	0.091	0.157	1.715	0.104	0.104	0.179	1.731	0.156	0.076	0.130	1.696	0.043
KSU-04	0.111	0.167	1.419	0.093	0.118	0.193	1.420	0.132	0.103	0.136	1.418	0.049
KSU-05	0.048	0.078	1.692	0.048	0.057	0.085	1.726	0.057	0.038	0.071	1.652	0.038
KSU-06	0.078	0.169	1.535	0.091	0.085	0.151	1.542	0.094	0.071	0.190	1.527	0.087
LisLb-01	0.720	0.942	1.237	0.591	0.722	0.920	1.349	0.684	0.717	0.967	1.109	0.484
TO-01	0.504	0.846	1.763	0.551	0.464	0.794	1.778	0.521	0.550	0.905	1.746	0.586
average	0.423	0.603	1.655	0.435	0.387	0.535	1.532	0.414	0.406	0.599	1.646	0.394

*akbl-02 did not submit single-type.

summarization task, 14 runs from 7 teams were submitted. For the classification task, 45 runs from 11 teams were submitted.

Table 5 lists the results of the segmentation task. The best recall was 1.000 of nami-11, the best precision was 0.940 of nami-01, and the best F-measure was 0.895 of RICT-01. Tables 6 and 7 list the quality question scores and the ROUGE scores, respectively. When the extra grade was regarded as incorrect, the best content score was 0.886 of nagoy-01. When the extra grade was regarded as correct, the best content score was 1.134 of KitAi-01. The best form score was 1.955 of KSU-01, and the best total score was 0.912 of KitAi-01. For ROUGE scores, nagoy-01 achieved the best scores, except in some cases. Table 8 lists the results of the classification task. The best accuracy was 0.942 of TTECH-07, -08 and -10. For support, the best recall was 0.731 of FU01-02; the best precision was 0.738 of KSU-03, -04, -07, and -08; and the best F-measure was 0.256 of TTECH-02. For against, the best recall was 1.000 of CUTKB-04, the best precision was 0.207 of TTECH-05, and the best F-measure was 0.216 of TTECH-05. For other, the best recall was 1.000 of TTECH-07, -08, -10, RICT-01, -05, -06, and STARS-01; the best precision was 0.947 of TTECH-02 and -05; and the best F-measure was 0.970 of TTHECH-07, -08, and -10.

6   Discussion

6.1   Segmentation Task

According to the system description, 20 out of 24 systems took a rule-based approach, such as clue expressions and/or heuristics, whereas the other 4 systems used machine learning. The best and average F-scores were 0.895 and 0.672 in

Table 7. ROUGE scores in formal run (all-topic)

		recall							F-measure						
		N1	N2	N3	N4	L	SU4	W1.2	N1	N2	N3	N4	L	SU4	W1.2
Surface Form	KitAi-01	0.440	0.185	0.121	0.085	0.375	0.217	0.179	0.357	0.147	0.096	0.067	0.299	0.168	0.188
	KitAi-02	0.390	0.174	0.113	0.078	0.320	0.200	0.154	0.343	0.154	0.101	0.069	0.281	0.173	0.176
	TTECH-01	0.278	0.060	0.035	0.020	0.216	0.092	0.096	0.240	0.055	0.031	0.018	0.187	0.079	0.111
	nagoy-01	0.459	0.200	0.131	0.089	0.394	0.229	0.186	0.361	0.151	0.097	0.064	0.305	0.169	0.192
	akbl-01	0.400	0.173	0.113	0.076	0.345	0.189	0.157	0.361	0.156	0.102	0.068	0.310	0.167	0.185
	akbl-02	0.326	0.124	0.080	0.057	0.269	0.147	0.112	0.320	0.119	0.077	0.055	0.262	0.141	0.144
	KSU-01	0.158	0.028	0.009	0.002	0.147	0.043	0.071	0.210	0.039	0.013	0.004	0.196	0.059	0.107
	KSU-02	0.185	0.043	0.021	0.014	0.167	0.063	0.080	0.230	0.056	0.027	0.017	0.209	0.080	0.116
	KSU-03	0.172	0.036	0.008	0.002	0.157	0.050	0.075	0.211	0.043	0.011	0.003	0.192	0.062	0.106
	KSU-04	0.171	0.044	0.013	0.002	0.153	0.055	0.072	0.219	0.056	0.017	0.003	0.195	0.072	0.106
	KSU-05	0.227	0.029	0.010	0.002	0.195	0.064	0.089	0.231	0.029	0.010	0.003	0.196	0.065	0.110
	KSU-06	0.221	0.038	0.013	0.004	0.187	0.065	0.086	0.230	0.038	0.012	0.004	0.192	0.067	0.108
	LisLb-01	0.251	0.120	0.079	0.058	0.211	0.132	0.103	0.226	0.107	0.071	0.051	0.188	0.115	0.118
	TO-01	0.267	0.093	0.061	0.045	0.230	0.117	0.105	0.272	0.086	0.052	0.036	0.233	0.110	0.133
Stem	KitAi-01	0.458	0.199	0.134	0.096	0.389	0.234	0.188	0.373	0.159	0.106	0.075	0.311	0.182	0.199
	KitAi-02	0.399	0.179	0.118	0.082	0.326	0.208	0.158	0.351	0.160	0.106	0.074	0.286	0.180	0.181
	TTECH-01	0.289	0.064	0.037	0.022	0.222	0.097	0.099	0.251	0.058	0.033	0.019	0.193	0.084	0.114
	nagoy-01	0.479	0.217	0.145	0.101	0.412	0.247	0.197	0.377	0.165	0.108	0.074	0.319	0.184	0.205
	akbl-01	0.415	0.184	0.122	0.083	0.357	0.203	0.164	0.375	0.165	0.110	0.074	0.322	0.179	0.195
	akbl-02	0.339	0.135	0.089	0.064	0.279	0.158	0.119	0.333	0.129	0.085	0.063	0.272	0.152	0.153
	KSU-01	0.161	0.028	0.010	0.002	0.148	0.044	0.071	0.214	0.040	0.013	0.004	0.197	0.061	0.108
	KSU-02	0.187	0.044	0.021	0.014	0.170	0.064	0.081	0.233	0.057	0.027	0.017	0.212	0.082	0.117
	KSU-03	0.175	0.036	0.008	0.002	0.159	0.052	0.075	0.217	0.044	0.011	0.003	0.196	0.065	0.108
	KSU-04	0.174	0.045	0.014	0.002	0.155	0.056	0.073	0.222	0.058	0.018	0.003	0.197	0.073	0.107
	KSU-05	0.230	0.029	0.010	0.002	0.199	0.066	0.090	0.236	0.030	0.010	0.003	0.201	0.067	0.112
	KSU-06	0.226	0.040	0.013	0.004	0.189	0.066	0.087	0.235	0.039	0.012	0.004	0.195	0.069	0.109
	LisLb-01	0.261	0.125	0.084	0.061	0.218	0.139	0.106	0.235	0.112	0.075	0.055	0.195	0.121	0.122
	TO-01	0.273	0.097	0.065	0.048	0.233	0.121	0.107	0.277	0.089	0.056	0.039	0.236	0.114	0.136
Content Word	KitAi-01	0.285	0.145	0.090	0.050	0.278	0.154	0.180	0.224	0.115	0.071	0.042	0.217	0.107	0.170
	KitAi-02	0.254	0.126	0.083	0.053	0.247	0.131	0.156	0.214	0.109	0.069	0.046	0.208	0.106	0.159
	TTECH-01	0.088	0.028	0.015	0.007	0.082	0.033	0.050	0.076	0.024	0.012	0.006	0.071	0.027	0.054
	nagoy-01	0.326	0.164	0.094	0.046	0.315	0.168	0.201	0.249	0.123	0.067	0.036	0.239	0.110	0.187
	akbl-01	0.256	0.113	0.065	0.034	0.247	0.124	0.148	0.224	0.098	0.056	0.031	0.216	0.100	0.158
	akbl-02	0.200	0.094	0.051	0.032	0.189	0.095	0.109	0.188	0.089	0.049	0.031	0.178	0.087	0.127
	KSU-01	0.048	0.001	0.000	0.000	0.047	0.007	0.032	0.059	0.001	0.000	0.000	0.058	0.009	0.043
	KSU-02	0.069	0.014	0.000	0.000	0.067	0.019	0.043	0.083	0.015	0.000	0.000	0.081	0.022	0.059
	KSU-03	0.041	0.002	0.000	0.000	0.041	0.007	0.027	0.050	0.002	0.000	0.000	0.050	0.008	0.036
	KSU-04	0.050	0.002	0.000	0.000	0.048	0.008	0.031	0.064	0.003	0.000	0.000	0.061	0.011	0.044
	KSU-05	0.067	0.002	0.000	0.000	0.062	0.013	0.041	0.063	0.003	0.000	0.000	0.057	0.011	0.043
	KSU-06	0.053	0.003	0.000	0.000	0.051	0.008	0.034	0.051	0.003	0.000	0.000	0.049	0.009	0.037
	LisLb-01	0.171	0.083	0.044	0.026	0.160	0.088	0.106	0.140	0.068	0.036	0.023	0.130	0.065	0.102
	TO-01	0.116	0.055	0.035	0.012	0.111	0.056	0.070	0.106	0.042	0.023	0.011	0.101	0.042	0.076

the rule-based systems, and those of the machine-learning systems were 0.872 and 0.815, respectively. The results were good overall, and the difference between the approaches was small.

6.2 Summarization Task

The aim is to obtain a score as close as possible to the total score of the all-topics section, which is obtained via human evaluation. Based on this premise, we examine how the scores, such as ROUGE, relate to the total score obtained by human evaluation. We would like to be able to evaluate the relative differences between the summaries of each system in this task, and we used Pearson's product moment correlation coefficient as an index to measure this relation. A score that has a stronger correlation with the total score of all-topics is considered to be a more appropriate index for the evaluation.

First, we investigated whether the result is influenced by the number of topics to be summarized. The correlation coefficient between the total score of all topics

Table 8. Result of classification task in formal run

	A	Support			Against			Other		
		R	P	F	R	P	F	R	P	F
FU01-01	0.624	0.417	0.057	0.100	0.076	0.041	0.053	0.648	0.938	0.766
FU01-02	0.373	<u>0.731</u>	0.057	0.106	0.183	0.045	0.072	0.362	0.943	0.523
FU01-03	0.909	0.089	0.164	0.115	0.008	0.020	0.011	0.970	0.936	0.953
FU02-01	0.842	0.027	0.040	0.032	0.095	0.033	0.049	0.899	0.933	0.916
FU02-02	0.840	0.073	0.063	0.068	0.069	0.030	0.042	0.895	0.933	0.914
TTECH-01	0.923	0.046	0.163	0.072	0.015	0.133	0.027	0.987	0.935	0.960
TTECH-02	0.896	0.260	0.252	<u>0.256</u>	0.221	0.199	0.209	0.943	<u>0.947</u>	0.945
TTECH-03	0.919	0.116	0.254	0.159	0.069	0.200	0.103	0.978	0.938	0.958
TTECH-04	0.921	0.043	0.134	0.065	0.015	0.133	0.027	0.985	0.934	0.959
TTECH-05	0.897	0.251	0.251	0.251	0.225	<u>0.207</u>	<u>0.216</u>	0.944	<u>0.947</u>	0.945
TTECH-06	0.918	0.132	0.269	0.177	0.080	0.206	0.115	0.976	0.939	0.957
TTECH-07	<u>0.942</u>	0.000	NaN	NaN	0.000	NaN	NaN	<u>1.000</u>	0.942	<u>0.970</u>
TTECH-08	<u>0.942</u>	0.000	NaN	NaN	0.000	NaN	NaN	<u>1.000</u>	0.942	<u>0.970</u>
TTECH-09	0.926	0.000	0.000	NaN	0.000	NaN	NaN	0.982	0.941	0.961
TTECH-10	<u>0.942</u>	0.000	NaN	NaN	0.000	NaN	NaN	<u>1.000</u>	0.942	<u>0.970</u>
akbl-01	0.923	0.118	0.344	0.176	0.034	0.097	0.050	0.983	0.939	0.960
ibrk-01	0.731	0.178	0.063	0.093	0.202	0.045	0.074	0.770	0.934	0.844
ibrk-02	0.731	0.178	0.063	0.093	0.202	0.045	0.074	0.770	0.934	0.844
RICT-01	0.933	0.000	NaN	NaN	0.000	NaN	NaN	<u>1.000</u>	0.933	0.965
RICT-02	0.932	0.002	0.091	0.004	0.004	0.111	0.008	0.998	0.933	0.964
RICT-03	0.893	0.118	0.145	0.130	0.111	0.117	0.114	0.949	0.940	0.944
RICT-04	0.894	0.114	0.143	0.127	0.111	0.117	0.114	0.950	0.939	0.944
RICT-05	0.933	0.000	NaN	NaN	0.000	0.000	NaN	<u>1.000</u>	0.933	0.965
RICT-06	0.933	0.000	NaN	NaN	0.000	NaN	NaN	<u>1.000</u>	0.933	0.965
RICT-07	0.932	0.084	0.440	0.141	0.042	0.407	0.076	0.994	0.937	0.965
STARS-01	0.933	0.000	NaN	NaN	0.000	NaN	NaN	<u>1.000</u>	0.933	0.965
STARS-02	0.889	0.002	0.002	0.002	0.000	NaN	NaN	0.953	0.933	0.943
STARS-03	0.889	0.002	0.002	0.002	0.000	NaN	NaN	0.953	0.933	0.943
STARS-04	0.889	0.002	0.002	0.002	0.000	NaN	NaN	0.953	0.933	0.943
tmcit-01	0.875	0.282	0.139	0.186	0.000	NaN	NaN	0.925	0.943	0.934
tmcit-02	0.893	0.239	0.160	0.192	0.000	NaN	NaN	0.946	0.942	0.944
tmcit-03	0.873	0.296	0.142	0.192	0.000	NaN	NaN	0.922	0.943	0.932
tmcit-04	0.879	0.319	0.161	0.214	0.000	NaN	NaN	0.928	0.944	0.936
tmcit-05	0.898	0.267	0.189	0.221	0.000	NaN	NaN	0.950	0.942	0.946
tmcit-06	0.878	0.292	0.148	0.196	0.000	NaN	NaN	0.927	0.943	0.935

(continued)

Table 8. (*continued*)

	A	Support			Against			Other		
		R	P	F	R	P	F	R	P	F
KSU-01	0.932	0.075	0.579	0.133	0.008	0.056	0.014	0.995	0.937	0.965
KSU-02	0.932	0.071	0.689	0.129	0.008	0.042	0.013	0.995	0.937	0.965
KSU-03	0.934	0.071	<u>0.738</u>	0.130	0.008	0.083	0.015	0.998	0.937	0.967
KSU-04	0.934	0.071	<u>0.738</u>	0.130	0.008	0.083	0.015	0.998	0.937	0.967
KSU-05	0.932	0.075	0.579	0.133	0.019	0.111	0.032	0.995	0.937	0.965
KSU-06	0.932	0.071	0.689	0.129	0.019	0.088	0.031	0.995	0.937	0.965
KSU-07	0.934	0.071	<u>0.738</u>	0.130	0.011	0.100	0.020	0.997	0.937	0.966
KSU-08	0.934	0.071	<u>0.738</u>	0.130	0.011	0.100	0.020	0.997	0.937	0.966
CUTKB-04	0.025	0.000	NaN	NaN	<u>1.000</u>	0.025	0.049	0.000	NaN	NaN
LisLb-01	0.914	0.021	0.065	0.032	0.037	0.080	0.051	0.976	0.935	0.955

and that of a single topic was 0.995, and the correlation coefficient between the total score of all topics and the total score of multiple topics was 0.991. However, the correlation with the single-topic is calculated using only 13 results, which do not include akbl-2 because akbl-2 was not submitted. Because both the correlation coefficients show a very strong positive correlation, we assume that the difference in the number of topics to be summarized has little influence on the evaluation. The following discussion will be based on the results of all topics.

Next, we investigated how the content and expression scores influence the total score. The correlation coefficient between the total score and the expression score was -0.046, which means it is almost uncorrelated. The correlation coefficient between the total score and the content score was obtained as follows. When the score of the extra grade X is 2 points (the same score as a correct answer, but corresponding to "the summary is different from *Togikai-dayori* but summarizes the contents of the original document"), the correlation coefficient is 0.983, whereas when the score of X is 0 points (the score for an incorrect answer), the correlation coefficient is 0.979, both of which indicate a very strong positive correlation. From this, we can infer that the content score influences the total score to a greater extent that the expression score. There are summaries other than *Togikai-dayori* that can be correct, but the comparison of the correlation coefficients (0.983 and 0.979) shows that the influence is small. Therefore, it seems that there is no problem in using *Togikai-dayori* as a correct answer.

Finally, we investigated which ROUGE approach is closest to human evaluation. Table 9 shows the correlation coefficient between the total score and each ROUGE score. Overall, it shows a very high correlation with any ROUGE score, but the highest value is the ROUGE-N4 0.972 (underlined in the table) based on the recall when using the morpheme sequence returned to the original form. As shown in Table 7, as the value of ROUGE-N increases, the absolute value of

Table 9. Correlation coefficient between the total score and each ROUGE score

	Recall							F-measure						
	N1	N2	N3	N4	L	W1.2	SU4	N1	N2	N3	N4	L	W1.2	SU4
Surface Form	0.924	0.955	0.964	0.968	0.915	0.953	0.893	0.900	0.942	0.957	0.959	0.852	0.946	0.882
Stem	0.928	0.959	0.968	<u>0.972</u>	0.918	0.956	0.900	0.912	0.950	0.965	0.968	0.866	0.954	0.894
Content Word	0.943	0.957	0.948	0.920	0.939	0.952	0.926	0.942	*0.963*	*0.953*	*0.924*	0.937	*0.956*	*0.935*

Table 10. Numbers in classifiers and encoders

classifier	num.	max	encoding	num.	max
Rule-Based	2	0.624	Key-Phrase	2	0.624
MaxEnt	1	0.909	One-Hot	19	0.942
3LP	2	0.842	Word Embedding	23	0.934
SVM	13	0.942	Unique	1	0.909
LSTM	13	0.934	Total	45	—
SVM+	7	0.932			
LSTM+	7	0.933			
Total	45	—			

the score approaches 0, so we referred mainly to -N1 (recall by content word) in previous QA Lab tasks. In this result, as well, the highest -N1 correlation (0.943) was obtained when we used the recall by the content word.

When comparing the recall and F-measure, the correlation was generally higher when the recall was used. However, when the content word was used, the F-measure showed higher correlation in -N2, -N3, -N4, -W1.2, and -SU4 (italicized in the table). Therefore, it might be possible to obtain a more appropriate evaluation by developing an evaluation method that considers whether functional expressions, such as modalities, are reproduced while also considering whether unnecessary content words are included.

6.3 Classification Task

We grouped the methods according to viewpoints that are shared by many methods, i.e., based on the type of machine-learning classifier and encoding.

Although most methods used a machine-learning classifier, there were two rule-based methods. Some methods employed a combination of classifiers, such as SVM and decision tree. Therefore, we determined the classifier groups as follows: rule-based, MaxEnt, three-layered perceptron (3LP), SVM, LSTM, a combination of SVM and other classifiers (SVM+), and a combination of LSTM and other classifiers (LSTM+). There was no method that used a combination of SVM and LSTM.

The encoding of the methods using the machine-learning classifier was performed through either one-hot encoding or word embedding. However, one method was observed to be an exception, as its encoding included folding a word and its location in a vector element. The rule-based classifiers used simple key phrases without encoding. Therefore, the encoding groups were determined as follows: key phrase, one-hot encoding, word embedding, and unique encoding. Table 10 lists the numbers of systems and the most accurate in the classifier and respective encoding groups.

The accuracy results of the machine-learning classifiers were observed to be better than those of the rule-based classifiers. The SVM classifier demonstrated the most accurate value of 0.942, whereas the LSTM classifier demonstrated a value of 0.934. The combinations of classifiers did not work as well as expected. An accuracy of 0.942 with one-hot encoding was the best, although it was only marginally higher than that of word embedding (0.934). Aker et al. [?] reported that the difference between the classifiers was marginal, and the results observed in this study exhibited a similar tendency.

While comparing the basic factors of classification with each other, it was observed that the results of fact-checkability were relatively low. As this is an important factor for a well-grounded argument, it may emerge as an issue in the future.

7   Conclusion

We described the overview of the NTCIR-14 QA Lab-PoliInfo task. Its goal is realizing complex real-world question answering (QA) technologies, to provide summaries of the opinions of assembly members and their reasons and conditions for such opinions, from Japanese regional assembly minutes. We conducted a dry run and a formal run, which include the segmentation, summarization, and classification tasks. Fifteen teams submitted 83 runs for the formal run. We described task description, collection, participation, and results.

References

1. Pariser, E.: The Filter Bubble: What the Internet is Hiding from You. Penguin Group, Westminster (2011)
2. Shibuki, H., et al.: Overview of the NTCIR-11 QA-lab task. In: Proceedings of the 11th NTCIR Conference (2014)
3. Shibuki, H., et al.: Overview of the NTCIR-12 QA lab-02 task. In: Proceedings of the 12th NTCIR Conference (2016)
4. Kimura, Y., et al.: Creating Japanese political corpus from local assembly minutes of 47 prefectures. In: Proceedings of Coling 2016 workshop, The 12th Workshop on Asian Language Resources, pp. 78–85 (2016)
5. Lin, C.-Y.: ROUGE: a package for automatic evaluation of summaries. In: Proceedings of the ACL-04 Workshop, vol. 8 (2004)

Short Text Conversation

Generating Topical and Emotional Responses Using Topic Attention

Zhanzhao Zhou, Maofu Liu$^{(\boxtimes)}$, Zhenlian Zhang, Yang Fu,
and Junyi Xiang

School of Computer Science and Technology,
Wuhan University of Science and Technology, Wuhan 430065, China
liumaofu@wust.edu.cn

Abstract. As an indispensable influencing factor of human-computer interaction experience, emotional cognitive behaviors in dialogues have aroused spread concern of researchers. However, existing emotional dialogue generation models tend to generate generic and universal responses. To address this problem, this paper proposes a topical and emotional chatting machine (TECM) that generates not only high-quality but also emotional responses. TECM utilizes the information obtained by the topic model as a prior knowledge to guide the generation of the responses, and the topic information is used as the input of the topic attention mechanism to improve the quality of responses. TECM also adopts a method of emotion category embedding to generate emotional responses. The empirical study on automatic evaluation metrics shows that TECM can generate diverse, informative and emotional responses.

Keywords: Emotional dialogue generation · Topic model · Topic attention mechanism · Emotion category embedding

1 Introduction

Conversation generation is a significant part of artificial intelligence, and it can enhance the human-computer interaction experience. There are two main methods of conversation generation: retrieval-based and generation-based. The former can only retrieve conversations in the conversation repository as responses, while the latter can generate new responses which never appear in the conversation repository. Therefore, the generation-based model has gradually become a research hotspot. Nevertheless, the generation-based dialogue system generates a large number of universal responses, and early researches focused on this problem.

In recent years, researchers tried to fuse emotions to the conversations. Expressing and understanding emotions and affects are not only human significant cognitive behaviors but also the key to enhancing human-computer interaction [1]. There are still many difficulties to be solved for constructing an emotional dialogue system. Firstly, the dialogue dataset with proper emotion tags are quite scarce. Secondly, how to evaluate the generated emotional responses is also difficult. To get a proper dataset where each instance is labelled with a user-specified emotion, the paper trains an emotion classifier to annotate the original dataset. The original dialogue dataset is provided by the NTCIR-14

© Springer Nature Switzerland AG 2019
M. P. Kato et al. (Eds.): NTCIR 2019, LNCS 11966, pp. 139–150, 2019.
https://doi.org/10.1007/978-3-030-36805-0_11

Chinese Emotional Conversation Generation task [2]. And the dataset consists of more than one million dialogues which are collected from Weibo, a popular social platform, where there are seas of short text conversations. For an emotional response, this paper evaluates it from two aspects simultaneously: emotion accuracy and content quality. Emotion accuracy is evaluated by the above emotion classifier which is used for original dataset annotation. And the dialogue quality evaluation metrics are present in Sect. 4.

Though a variety of models have been proposed for conversation generation from large-scale social data, it is still quite challenging (and yet to be addressed) to generate emotional responses. Generated responses are indeed emotional yet they are usually generic and universal. These models often only address the emotion factors in the responses while ignoring the quality of the dialogues. The large number of generic and universal responses makes it difficult to detect the emotion of the generated response. To deal with these problems, this paper absorbs the aforesaid topic attention mechanism, and presents a Topical and Emotional Chatting Machine (TECM), which can improve the quality of responses and put emotions to responses simultaneously. The TECM model is based on seq2seq model [3], and leverages the information of the topic model and emotion category embedding to guide the generation of dialogues. The generated conversations are not only informative but also emotional in this way. In general, the contribution of this paper is mainly to add the topic information to the emotional dialogue system to generate more informative and diverse responses.

Section 2 describes the related work of the dialogue systems. Section 3 details on the methods based on the TECM model. In Sect. 4, the dataset preparation process and the experimental results are discussed. The Sect. 5 concludes the paper and discusses the future works.

2   Related Work

Dialogue generation systems are divided into the retrieval-based systems and the generation-based systems. Ji et al. proposed a retrieval-based model [4] that combines multiple short text matching approaches and response ranking techniques. Shang et al. applied the seq2seq model [5] to the short text conversation generation task. Cao et al. presented a latent variable dialogue model [6], where the latent variable is like a topic. Although the generation-based dialogue system can create new responses, researchers have gradually found that the generation-based systems suffer from severe universal response generation (e.g. *me too*). Li et al. proposed the MMI model [7] that regards the maximum mutual information as an objective function to alleviate this problem. Yao et al. proposed a model [8] that incorporated inverse document frequency (IDF) to measure the differences between the generated responses. Xing et al. presented TA-Seq2Seq model [9] that utilized topic information to guide the generation of responses. Recently, the researchers did not satisfy the rigid dialogues, and they began to incorporate emotions to guide the generation of emotional responses. The Emotional Chatting Machine (ECM) model, proposed by Huang et al. [10], can generate responses appropriate not only in content but also in emotion, but the ECM need to be given a user-specified emotion class instead of deciding the most appropriate emotion category for the response. The Affect-LM model proposed by Ghosh et al. [11] can generate expressive text at varying degrees of emotion strength without affecting grammatical correctness. Asghar et al. [12] proposed a model which can produce more natural and emotionally rich responses without user-specified the emotion class.

These works are either for the quality of the responses or for the emotions of the responses. The TECM model takes different ways to improve both the quality and emotional expression of the conversation particularly. The proposed model incorporates the information of the topic model to enhance the diversity of dialogue, and adopts the emotion category embedding to make responses emotional. The experimental results show that the performance of TECM model combining the above approaches outperforms ECM model.

3   Methods

Given a post $X = \{x_1, x_2, x_3, \ldots, x_N\}$ and a user-specified emotion tag e, the task requires the generation of $Y = \{y_1, y_2, y_3, \ldots, y_M\}$ that is coherent with the post and in accordance with the user-specified emotion e. Emotion tags include like, sadness, disgust, anger, happiness, and other. The entire dataset will be covered in detail in Sect. 4.

To deal with this task, this paper presents a TECM model. As is shown in Fig. 1, the TECM model is a seq2seq model with the message attention mechanism, the topic attention mechanism and the emotion category embedding mechanism. The message attention mechanism mainly focus on the post. The paper takes the topic information as the input of the topic attention mechanism. The TECM model uses the topic information obtained by the topic model as a prior knowledge to generate an informative response. For the user-specified emotion requirement, The TECM model employs emotion category embedding to satisfy it.

3.1   Seq2seq with Attention Model

The TECM is based on a seq2seq model with the gated recurrent units (GRU) [13]. The GRU unit is a variant of the long-short term memory (LSTM) [14] unit. The structure of GRU unit is simpler. The model with GRU units spends less time than the one with LSTM units, whereas their effects are similar. The seq 2seq model is an encoder-decoder framework which contains an encoder reading source texts and a decoder generating target texts. In dialogue generation task the source text is the post of a dialogue and the target text is the response of a dialogue. The encoder reads the words of the input sequence one by one and then encodes them into the context vector which is referred to as the message context vector above. During the decoding phase, the predicted tokens are decoded by the decoder one by one according to the context vector and the target text. Given a post $X = \{x_1, x_2, x_3, \ldots, x_N\}$, the encoder converts X to a sequence of hidden vectors $h = \{h_1, h_2, h_3, \ldots, h_N\}$. This process can be defined as follows:

$$h_t = GRU(h_{t-1}, x_t) \tag{1}$$

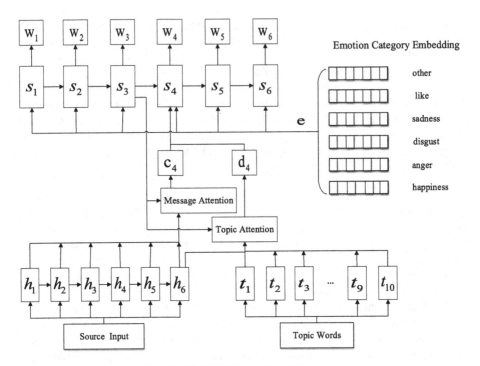

Fig. 1. The TECM model architecture.

The specific implementation of the GRU unit is described as follows:

$$z_t = \sigma(W_z \cdot [h_{t-1}, x_t]) \tag{2}$$

$$r_t = \sigma(W_r \cdot [h_{t-1}, x_t]) \tag{3}$$

$$s_t = tanh(W_h \cdot [r_t \circ h_{t-1}, x_t]) \tag{4}$$

$$h_t = (1 - z_t) \circ s_t + z_t \circ h_{t-1} \tag{5}$$

where the operator $\circ$ represents the dot product between two vectors.

By quickly scanning the image, human being obtains the target area that needs to be focused on, which is the attention focus. The target area will acquire more attention resources, while other areas will be ignored. The proposed attention mechanism [15] can dynamically attend on key information of the input post at each decoding step. Thus, the context vector c_t is a weighted sum of the hidden states, which is defined as follows:

$$c_t = \sum_{j=1}^{N} \alpha_{tj} h_j \tag{6}$$

where the weight α_{tj} is a measure of how much the attention mechanism acquire from each hidden representation h_j. This formula is described as follows:

$$\alpha_{tj} = \frac{\exp\left(\eta\left(s_{t-1}, h_j\right)\right)}{\sum_{k=1}^{N} \exp(\eta(s_{t-1}, h_k))} \tag{7}$$

where η denotes the Multi-Layer Perception.

In the decoding phase, the context vector, the previous hidden state and the target text will be regarded as the inputs of GRU units to compute the hidden state of the next state.

$$s_t = GRU(s_{t-1}, y_{t-1}, c_t) \tag{8}$$

The context vector in the decoding phase is different at each state due to the attention mechanism. Finally, the target function of seq2seq can be written as:

$$y_t \sim p_i = p(y_t | x_1, x_2, x_3, \ldots, x_N, y_1, y_2, \ldots, y_{t-1}, c_t) \tag{9}$$

$$= softmax(W s_t) \tag{10}$$

According to this formula, the next token is decoded from the source text, the previously decoded tokens and the context vector.

3.2   Topic Model

The TECM model utilizes the information from the topic model as the prior knowledge to make the responses more diverse. The TECM employs a topic attention mechanism to enable the incorporation of topic information. First, the topic model generates a quantity of topic words for each post, and the topic attention mechanism combines the additional topic information with the source texts to generate informative responses.

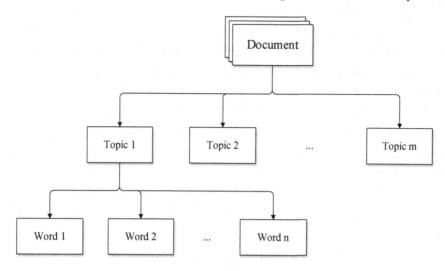

Fig. 2. NMF model architecture.

The topic model is a statistical model in natural language processing that discovers potential topics for a document. There are some popular topic models such as LDA [16], NMF [17] and so on. The first one is mainly used for topic extraction of long texts, but the experimental data in this paper is the microblog short text. NMF model is faster than the LDA. Hence, this paper leverages the NMF model. NMF model holds the view that each document has several topics, and each topic is composed of several topic words. The structure of the NMF model is shown in Fig. 2.

NMF is a decomposition method that decomposes a matrix into two non-negative matrices. The mathematical formula is defined as follows:

$$V_{m \times n} \approx W_{m \times k} \times H_{k \times n} = U_{m \times n} \tag{11}$$

where $V_{m \times n}$ represents the original matrix to be decomposed, and $U_{m \times n}$ is expressed as the product of $W_{m \times k}$ and $H_{k \times n}$, which should be approximately equal to the original matrix.

3.3   Topic Attention Mechanism

NMF model will generate m (m = 100) topic probabilities for each conversation in the training dataset. The topic of maximum probability will be selected. Then, the top n (n = 10) topic words with the high probability under this topic will be selected as the topic information of the conversation. These n topic words will be regarded as the inputs of the topic attention mechanism. A post with 10 topic words are shown in Example 1. For this post, it's related to the topic of cold. It can be seen that the topic model can assign proper topic words to a post. Most of the topic words relate to the topic of cold. But some words are irrelevant or even contrary to the topic. A variety of topic models will be used in the future work.

Example 1:
Post:　感冒了, 好困, 今天请假好了!
(Cold! Sleepy! I'll ask for leave today.)
Topic Words:　生病 特别 感冒 难受 天气 身体 舒服 日子 肚子 痛苦
(sick, special, cold, uncomfortable, weather, body, comfortable, day, belly, pain)

Because of the inspiration of the preceding TA-Seq2Seq model, this paper integrates prior topic information into the model through attention mechanism. For each pair of <post, response>, they are assigned to 10 topic words. Given a topic word sequence $T = \{t_1, t_2, t_3, \ldots, t_{10}\}$, the topic attention mechanism takes as the input the topic word embedding. The T sequence will eventually be transformed into a topic context vector (similar to c_t in Eq. (6)). And the topic attention calculation process is described as follows:

$$\beta_{tj} = \frac{\exp\left(\eta\left(s_{t-1}, t_j, h_N\right)\right)}{\sum_{k=1}^{10} \exp(\eta(s_{t-1}, t_k, h_N))} \tag{12}$$

$$d_t = \sum_{j=1}^{10} \beta_{tj} t_j \tag{13}$$

where h_N is the last hidden representation of the source text, and the topic attention mechanism leverages h_N to focus on topic words that are more relevant to the source text. Finally, the topic context vector d_t and the context vector c_t will be the inputs of GRU units.

3.4 Emotion Category Embedding

TECM is required to generate responses with user-specified emotions. Inspired by ECM proposed by Huang, this paper utilizes the emotion category embedding to make responses emotional. There are six target emotion categories: like, sadness, anger, disgust, happiness and other. This paper holds the hypothesis that the emotional factors in emotional response generation is controlled by an abstract emotion category vector. In TECM model, each abstract emotion category vector can be viewed as a mechanism to activate the emotional factors in emotional response generation. In the specific implementation, each emotion category can be expressed as a 100-dimensional real-valued vector e. Emotion category e will be given a random initial value at the beginning and can be learned through the training process. This process is defined as follows:

$$s_t = GRU(s_{t-1}, y_{t-1}, c_t, d_t, e) \tag{14}$$

This approach mainly changes the formula (8) of decoding, the input in each decoding step is concatenated with a goal emotion vector, the message context vector and the topic context vector.

4 Experiments

4.1 Experimental Settings

Firstly, this paper trains an emotion classifier with the accuracy of 92.6% by BERT [18] model. BERT is a language model developed by the Google team and has achieved the best performances in 11 different NLP tests. The training dataset of the model comes from the emotion classification tasks of NLPCC 2013 and NLPCC 2014. This dataset contains 38,122 short microblog texts which are annotated with emotions manually. Secondly, the original dataset used for dialogue generation task will be annotated by the above-described emotion classifier. The conversation dataset consists of 1,682,140 instances. This paper filters out dialogues in which the number of Chinese characters accounts for less than one third of the total number of characters. In this way, pure English dialogues and dialogues with too few Chinese characters are removed. 200 instances for each emotion category are selected randomly to be a part of test dataset. Thus, 1,200 instances compose the final test dataset. A validation dataset is acquired by the same way. The rest of the original dataset, 1,679,740 instances, make up the training dataset.

The TECM model is implemented by Tensorflow[1] framework. This model is a two-layer GRU structure whose hidden layer size is 256. Word embedding vectors are pre-trained 256-dimensional vectors. Pre-trained word vectors can not only improve the performances of models but also shorten the training time greatly. The vocabulary size in the model is set to 4,097. The dimension of the emotion category embedding vector is 100. The total number of topics in the NMF model is 100. The top 10 topic words of the selected topic will be the extra information of the dialogue. The batch size to use during training is 128. The initial learning rate is 0.99 and decays by 0.99. This paper adopts stochastic gradient descent (SGD) algorithm.

How to evaluate the quality of dialogues generated by the dialogue system is a key point in the research, which is mainly divided into automatic evaluation and manual evaluation. It takes too much time to evaluate the total results by the manual evaluation. There are some automatic evaluation metrics such as BLEU [19] and ROUGE [20] in this area. Liu et al. holds that BLEU metrics have positive correlation on the chitchat oriented Twitter dataset [21]. BLEU is a popular machine translation evaluation metric, which is provided for analyzing the co-occurrence times of n-tuples between candidate translations and reference translations.

Perlexity is a metric of language model. It is used for evaluating the quality of language model. That is to say, it is provided for seeing whether a sentence is smooth. Distinct-1 and Distinct-2 are proposed to measure the diversity between responses. Distinct-1 and Distinct-2 describe the richness of vocabulary. The higher the scores are on these metrics, the more informative the generated dialogues are.

For comparison with the TECM model, this paper regards the following models as baselines.

S2SE: the standard seq2seq model with attention and emotion category embedding.
ECM: the Emotional Chatting Machine proposed by Huang et al.

4.2 Experimental Results

We participated in the NTCIR-14 competition. The organizer of the contest adopts the method of mannual evaluation. If a generated response is fluent and coherent with the post, you can get 1 point, otherwise you can not get scores. And if the generated response is also in accordance with the user-specified emotion, you can get another 1 point. Our WUST system scores 587 in Overall Score. And for the happiness class, WUST system generates many appropriate responses and ranks the top in contrast to other emotion classes. It is principally because the corresponding training dataset is enormous. Table 1 shows the official evaluation results of our system.

Besides the official evaluation, we have made some additional experiments. Due to the lack of manual evaluation resources, we construct our own test dataset and evaluate it on the automatic evaluation metrics. There are two tables to show the results of the experiments. Table 2 shows the accuracies of the models in each emotion class. In particular, instances with *other* label will not be evaluated. Table 3 shows the results of

---

[1] https://github.com/tensorflow/tensorflow.

Table 1. The official evaluation results

Submissions/Emotions	Label0	Label1	Label2	Total	Overall score	Average score
WUST	601	211	188	1000	587	0.587
Like	117	36	47	200	130	0.65
Sadness	124	31	45	200	121	0.605
Disgust	111	69	20	200	109	0.545
Anger	137	48	15	200	78	0.39
Happiness	112	27	61	200	149	0.745

the models on metrics about the dialogue quality. From these two tables, the TECM model outperforms other baselines both on emotion metrics and dialogue quality metrics.

ECM proposed three mechanisms to incorporate emotions in the response generation, while TECM only levarages one of the three emotion mechanisms to guide the emotional response generation. In spite of one emotion mechanism, the TECM model has a slightly higher score on emotion metrics than the ECM model. It is because that TECM absorbs the topic attention mechanism to improve the dialogue quality. This suggests that high-quality responses can contribute to the emotional response generation. The score of S2SE on the happiness accuracy metric is the highest among the models. Because this model tends to generate universal emotional responses (e.g. *haha*) regardless of the input.

Table 2. Accuracies of each emotion for three models.

Model	Like	Sadness	Disgust	Anger	Happiness	Overall
S2SE	0.885	0.775	0.565	0.725	**0.940**	0.778
ECM	0.910	0.805	0.520	**0.805**	0.925	0.793
TECM	**0.920**	**0.815**	**0.560**	0.780	0.925	**0.800**

On the one hand, the performance of TECM model is the best on the perplexity metric. It shows that the TECM model can generate emotional responses without affecting the grammatical structure of sentences. On the other hand, the TECM model outperforms ECM model with a BLEU score increase of 10%. This means that there are more identical words between the generated responses and the gold responses. Finally, the S2SE model scores poor marks on Distinct-1 and Distinct-2 metrics, which means that S2SE model has a tendency to generate generic and universal responses. The scores on Distinct-1 and Distinct-2 metrics of the TECM model are much higher than the ECM baseline. The big increases of scores on Distinct-1 and Distinct-2 metrics are 10% and 18% respectively. It indicates that prior topic information makes the generated responses more diverse and informative.

Table 3. Performances of models on several automatic evaluations.

Model	Perplexity	BLEU	Distinct-1	Distinct-2	Emotion accuracy
S2SE	47.20	7.25%	0.0757	0.2610	0.778
ECM	54.97	7.09%	0.0825	0.2764	0.793
TECM	**40.42**	**7.80%**	**0.0911**	**0.3271**	**0.800**

Table 4. Examples of models.

Model	Emotion	Post	Response
ECM	like	你好喜欢玩网游喔！ You like to play online games very much!	我喜欢玩游戏 I like playing games.
	sad	看了我好饿。。。 I'm hungry…	我也饿了。。。 I'm hungry too…
	disgust	你这失眠的娃儿！ You insomniac baby!	我也失眠，郁闷死了 Me too. I'm insomniac too. I'm depressed.
	angry	哼，找骂啊！！ Hum, look for curses!!	我骂你了吗？ Did I scold you?
	happy	幸福来的太突然[哈哈] Happiness comes too suddenly, hahaha.	幸福的孩子 A lucky baby.
TECM	like	我的排骨很好吃呀 My spareribs are delicious.	我喜欢吃排骨 I like to eat spareribs.
	sad	来吧！孩纸！ Come on! Baby!	我已经累趴下了！ I'm tired!
	disgust	突然很烦!想死的心都有！ Suddenly very annoyed! Want to die!	烦死你！ Pester you to death.
	angry	今天真够生气的了 I'm really angry today.	消消气 Take it easy.
	happy	哎都知道我是吃货了哈哈 Well, everyone knows I'm a chowhound, haha.	哈哈，吃货 Haha, a chowhound.

There are also some examples in Table 4 for ECM model and TECM model. From these examples, the TECM model makes responses more diverse and informative, and the emotions contained in the generated responses are consistent with the goal emotions.

5 Conclusions

This paper proposes a TECM model, which is used to solve the problem that responses generated in emotional dialogue system are not informative enough. Hence, this paper incorporates additional topic information to guide the generation of responses. The topic information of the dialogue comes from a NMF model. The topic attention mechanism absorbs extra topic messages to reduce the generic response generation probability. To generate emotional responses, the TECM model leverages the emotion category embedding. The experimental results show that the conversations generated by TECM model are not only more diverse and informative, but also emotional obviously.

Nevertheless, the correlation between topic information and dialogues is poor or even irrelevant. In the future work, more various topic models will be used for extracting better topic information, thus generating more diverse and informative responses for the dialogue system.

Acknowledgments. The work presented in this paper is partially supported by the Major Projects of National Social Foundation of China under Grant No. 11&ZD189.

References

1. Prendinger, H., Mori, J., Ishizuka, M.: Using human physiology to evaluate subtle expressivity of a virtual quizmaster in a mathematical game. Int. J. Hum.-Comput. Stud. **62**, 231–245 (2005)
2. Zhang, Y., Huang, M.: Overview of the NTCIR-14 short text generation subtask: emotion generation challenge. In: Proceedings of the 14th NTCIR Conference on Evaluation of Information Access Technologies (2019)
3. Sutskever, I., Vinyals, O., Le, Q.V.: Sequence to sequence learning with neural networks. In: Advances in Neural Information Processing Systems, pp. 3104–3112 (2014)
4. Ji, Z., Lu, Z., Li, H.: An Information Retrieval Approach to Short Text Conversation. arXiv preprint arXiv:1408.6988 (2014)
5. Shang, L., Lu, Z., Li, H.: Neural responding machine for short-text conversation. In: ACL, pp. 1577–1586 (2015)
6. Serban, I.V., Sordoni, A., Lowe, R., et al.: A hierarchical latent variable encoder-decoder model for generating dialogues. In: Thirty-First AAAI Conference on Artificial Intelligence. (2017)
7. Li, J., Galley, M., Brockett, C., et al.: A diversity-promoting objective function for neural conversation models. In: ACL, pp. 110–119 (2016)
8. Yao, K., Peng, B., Zweig, G., et al.: An Attentional Neural Conversation Model with Improved Specificity. arXiv preprint arXiv:1606.01292 (2016)

9. Xing, C., Wu, W., Wu, Y., et al.: Topic aware neural response generation. In: Thirty-First AAAI Conference on Artificial Intelligence (2017)

10. Zhou, H., Huang, M., Zhang, T., et al.: Emotional chatting machine: emotional conversation generation with internal and external memory. In: Thirty-Second AAAI Conference on Artificial Intelligence (2018)

11. Ghosh, S., Chollet, M., Laksana, E., et al.: Affect-LM: a neural language model for customizable affective text generation. In: ACL, pp. 634–642 (2017)

12. Asghar, N., Poupart, P., Hoey, J., Jiang, X., Mou, L.: Affective neural response generation. In: Pasi, G., Piwowarski, B., Azzopardi, L., Hanbury, A. (eds.) ECIR 2018. LNCS, vol. 10772, pp. 154–166. Springer, Cham (2018). https://doi.org/10.1007/978-3-319-76941-7_12

13. Chung, J., Gulcehre, C., Cho, K.H., et al.: Empirical evaluation of gated recurrent neural networks on sequence modeling. arXiv preprint arXiv:1412.3555 (2014)

14. Hochreiter, S., Schmidhuber, J.: Long short-term memory. Neural Comput. **9**, 1735–1780 (1997)

15. Luong, M.T., Pham, H., Manning, C.D.: Effective Approaches to attention-based neural machine translation. In: EMNLP, pp. 1412–1421 (2015)

16. Blei, D.M., Ng, A.Y., Jordan, M.I.: Latent Dirichlet allocation. J. Mach. Learn. Res. **3**, 993–1022 (2003)

17. Lee, D.D., Seung, H.S.: Algorithms for non-negative matrix factorization. In: Advances in neural information processing systems, pp. 556–562 (2001)

18. Devlin, J., Chang, M.W., Lee, K., et al.: BERT: pre-training of deep bidirectional transformers for language understanding. In: ACL, pp. 4171–4186 (2018)

19. Papineni, K., Roukos, S., Ward, T., et al.: BLEU: a method for automatic evaluation of machine translation. In: Proceedings of the 40th annual meeting on association for computational linguistics. Association for Computational Linguistics, pp. 311–318 (2002)

20. Lin, C.Y.: Rouge: a package for automatic evaluation of summaries. In: Text Summarization Branches Out, pp. 74–81 (2004)

21. Liu, C.W., Lowe, R., Serban, I.V., et al.: How NOT to evaluate your dialogue system: an empirical study of unsupervised evaluation metrics for dialogue response generation. In: ACL, pp. 2122–2132 (2016)

A Hybrid Framework of Emotion-Aware Seq2Seq Model for Emotional Conversation Generation

Xiaohe Li, Jiaqing Liu, Weihao Zheng, Xiangbo Wang, Yutao Zhu, and Zhicheng Dou[✉]

School of Information, Renmin University of China, Beijing, China
{lixiaohe,jiaqingliu,zheng-weihao,xiangbo28,ytzhu,dou}@ruc.edu.cn

Abstract. This paper describes RUCIR's system in NTCIR-14 Short Text Conversation (STC) Chinese Emotional Conversation Generation (CECG) subtask. In our system, we use the Attention-based Sequence-to-Sequence (Seq2Seq) method as our basic structure to generate emotional responses. This paper introduces (1) an emotion-aware Seq2Seq model and (2) several features to boost the performance of emotion consistency. Official results show that our model performs the best in terms of the overall results across the five given emotion categories.

Keywords: Emotional Conversation Generation · Sequence to sequence model · Attention mechanism · Copy mechanism

1 Introduction

The human-computer conversation is one of the most challenging tasks in natural language processing (NLP). Particularly, short text conversation (STC) which simulates human real-life dialogues has attracted more and more attention.

STC can be defined as a kind of single-turn conversation formed by two short texts, with the initial utterance given by a human user and the response given by a computer. STC task (STC-1) is first proposed in NTCIR-12 [8], which was taken as an information retrieval (IR) problem and aimed to retrieve an appropriate response in the repository to reply a user-issued utterance. At NTCIR-13 [7], STC-2 encouraged the participants to combine retrieval-based methods and generation-based methods to make a response for a new user-issued utterance. This year, we participated in NTCIR-14 STC-3 CECG subtask [13]. Compared with the former tasks, CECG aims at generating emotional Chinese responses that are not only reasonable in content but also consistent with a given emotion. The pre-defined emotion categories include *like, sad, disgust, anger, happy* and *other*.

In general, conversation systems can be categorized into retrieval-based and generation-based. Retrieval-based methods maintain a large repository of conversation data and consider the user-issued utterance as a query, then return

© Springer Nature Switzerland AG 2019
M. P. Kato et al. (Eds.): NTCIR 2019, LNCS 11966, pp. 151–162, 2019.
https://doi.org/10.1007/978-3-030-36805-0_12

a most proper response through information retrieval techniques. Generation-based methods generate responses with natural language generation models learned from the conversation data. A typical generation method is the sequence-to-sequence (Seq2Seq) neural network model [4,6,10,11]. The Seq2Seq model generally incorporates an encoder and a decoder. The encoder is used to represent the input message as a vector, based on which the decoder generates a new response. The encoder and the decoder are usually constructed by recurrent neural networks (RNNs). Since the structure of RNN is naturally suitable to model time-series data, the Seq2Seq model can capture semantic and syntactic relations between user-issued utterances and responses. An attention mechanism is often used to enhance the model on learning patterns from data [1,5].

In this work, we use the Seq2Seq model with attention mechanism as our basic model to build the conversation system. As shown in Fig. 1, our system consists of four modules. The first one is a rule-based template in which important information such as entities, weather and other keywords are taken into account. The second module comprises multiple fine-tuned Seq2Seq models to generate responses in different emotions respectively. The third module is a single emotion-aware Seq2Seq model with the input of emotion factors and emotion keywords. Finally, a re-ranker is designed to select the final responses.

The rest of paper is organized as follows: We will introduce our system architecture in detail at first. Then we report the experimental results in Sect. 3. Finally, we will make a brief conclusion of our work.

2   System Architecture

As shown in Fig. 1, in our model architecture, we use three different methods to generate responses, and then use re-rank to select the best response. We will then describe them in detail.

2.1   Data Pre-processing

Good quality of training data is essential for training a good model. We first process the dataset and remove the noisy information that is useless or even harmful to the model training.

Token-Level Data Pre-processing. We artificially check the data and summarize some patterns for meaningless responses. More specifically, we first identify the responses that contain: (1) emoji and kaomoji, (2) dialect and online buzzwords, (3) repeated expressions in word level and sentence level, (4) meaningless beginnings of sentence such as "Yes" and "Haha", (5) mention and repost characters ('@' or '//@'). We filter out these meaningless expressions and emotion icons in the original posts or responses, and replace the dialect and buzzwords with Mandarin based on the dictionary.

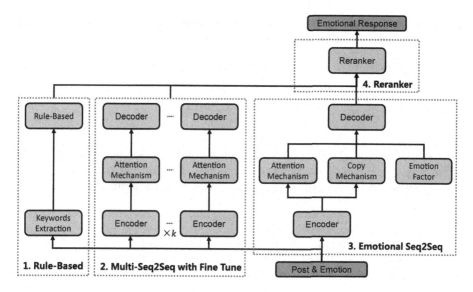

Fig. 1. The structure of our system

Sentence-Level Data Pre-processing. We retain the word segmentation of the original dataset and filter out post-response pairs that are not Chinese or too short (the post or response with less than three characters).

Since the Seq2Seq model tends to generate trivial and meaningless responses which appear many times in dataset such as "Haha" and "What's up?", we remove sentences that occur more than 100 times and simplify tokens that continuously and repeatedly appear more than twice. For example, "What's up?" appears 4,014 times in responses, thus all post-response pairs with such a response are removed from the dataset. And "Hahahahaha" is simplified as "Haha".

2.2 Emotion-Aware Seq2Seq Model

We use an emotion-aware Seq2Seq model to generate different responses with different emotion categories, which is inspired by the ECM model [14]. Based on the Seq2Seq model with attention mechanism, we introduce emotion factors by emotion embeddings, and increase the probability of generating emotional words by the copy mechanism.

Seq2Seq Model is originally proposed for machine translation [10]. Then Shang et al. applied this model into neural response generation [6]. After that, tremendous approaches have been proposed for response generation based on the Seq2Seq model [4,11]. In this work, we also build our model based on it.

In general, the Seq2Seq model consists of an encoder and a decoder. Both of them can be implemented with an RNN and its variations such as long-short term memory (LSTM) [3] and gated recurrent unit (GRU) [2]. We use the GRU in this work, which can be formulated as:

$$
\begin{aligned}
z &= \sigma(W_z x_t + U_z h_{t-1}), \\
r &= \sigma(W_r x_t + U_r h_{t-1}), \\
s &= \tanh(W_s x_t + U_s(h_{t-1} \circ r)), \\
h_t &= (1 - z) \circ s + z \circ h_{t-1}.
\end{aligned}
\tag{1}
$$

Assume $\mathbf{x} = (x_1, x_2, \cdots, x_n)$ is a sequence of input post containing n words, and $\mathbf{y} = (y_1, y_2, \cdots, y_m)$ is a generated response. The encoder transforms $\mathbf{x}$ into a sequence of hidden states $h = (h_1, h_2, \cdots, h_n)$, which is defined as:

$$
h_t = \mathrm{GRU}_{\mathrm{encoder}}(x_t, h_{t-1}),
\tag{2}
$$

where h_t is the hidden state of the encoder at time step t.

The decoder is another GRU maximizing the conditional probability of a target word y_t, which can be formulated as:

$$
p(y_t | \{y_1, y_2, \cdots, y_{t-1}; \mathbf{x}\}) = p(y_t | s_t) = \mathrm{softmax}(W_o s_t),
\tag{3}
$$

$$
s_t = \mathrm{GRU}_{\mathrm{decoder}}(y_{t-1}, s_{t-1}),
\tag{4}
$$

where s_t is the hidden state of the decoder at time t, the y_0 is the start of sentence (SOS) token and s_0 is equal to h_n.

Attention Mechanism is often used to improve the model on learning patterns from the data [1,5]. In a vanilla Seq2Seq model, the decoder generates the next word y_t only depends on the word y_{t-1} and the s_{t-1} at time step t. Since s_0 is equal to the final encoder hidden state h_n, the useful information of words in the front part of the source sequence is neglected.

Besides, at different decoding steps, vanilla Seq2Seq model can not measure the importance of different words in the source sequence. On the contrary, in the attention mechanism, each word y_t corresponds to a context vector c_t calculated by a weighted sum of the encoder hidden states h. In this work, we use the Luong attention mechanism [5], which can be formulated as:

$$
s_t = \mathrm{GRU}_{\mathrm{decoder}}(y_{t-1}, s_{t-1}, c_t),
\tag{5}
$$

$$
c_i = \sum_{j=1}^{n} \alpha_{ij} h_j,
\tag{6}
$$

$$
\alpha_{ij} = \frac{\exp(e_{ij})}{\sum_{k=1}^{n} \exp(e_{ik})},
\tag{7}
$$

$$
e_{ij} = s_{i-1}^{\mathrm{T}} W_a h_j.
\tag{8}
$$

Emotion Factor. We concatenate the emotion category embedding as the emotion factor to each decoding step, which can introduce additional emotion information when generating a response with a given emotion. Therefore the decoder can generate responses more emotional under the given emotion while predicting the next word. The embedding of every emotion category is initialized randomly, and then learned in the training process. Each emotion factor is represented by a real-valued and dense vector. With the emotion factor, the hidden state of the decoder can be updated as:

$$s_t = \text{GRU}_{\text{decoder}}(y_{t-1}, s_{t-1}, c_t, e_i), \tag{9}$$

where e_i is the emotion embedding of the specific emotion category i.

Copy Mechanism. Intuitively, emotion expressions usually have some distinct emotion words. For example, "I lost sleep last night." and "I was so sad about insomnia last night.". Both of them express sadness, but the former one seems to describe a fact, while the latter expresses sadness directly. Therefore, we can divide emotion expressions into implicit expressions and explicit expressions. Apparently, an emotion word is so expressive that can be easily perceived and recognized by humans. Since we prefer explicit expressions, we want to increase the probability of generating emotion words.

Zhou et al. used a type selector which can operate the distribution of generic and emotion words to control the generation of emotion words [14]. Song et al. applied a copy mechanism to enrich the useful and informative words in conversation system [9]. Inspired by these works, we use the copy mechanism to increase the probability of emotional words in the generation process, and make the emotional expressions of generated responses more explicit.

We first build an emotion dictionary for different emotions by clustering the tokens from responses. We call the words in the emotion dictionary as emotional words, and call the other words as non-emotional words. For non-emotional word, we just use the original probability. For emotional word, we add the copy probability to original probability to increase the generating probability. The final generated word probability distribution is given by:[1]

$$p(y_t|s_t) = p_{\text{ori}}(y_t|s_t) + p_{\text{emo}}(y_t|s_t, E), \tag{10}$$

$$p_{emo}(y_t|s_t, E) = \text{softmax}(EW_e s_t), \tag{11}$$

where p_{ori} is the original probability distribution, p_{emo} is the additional emotion words distribution. If y_t is not an emotional word, the corresponding probability p_{emo} would be zero. E is the word embeddings of words in emotion dictionary. W_e are the parameters for matching E and s_t. The composition of emotion dictionary will be introduced in the Sect. 2.5.

2.3   Multi-Seq2Seq Models with Fine Tune

The previous part of our system uses only one Seq2Seq model. In this part, we train different Seq2Seq models for different emotions. To generate responses

---

[1] Here is just for the convenience of explanation, because the sum of two probability may be greater than 1. Actually, we will guarantee the value is not greater than 1.

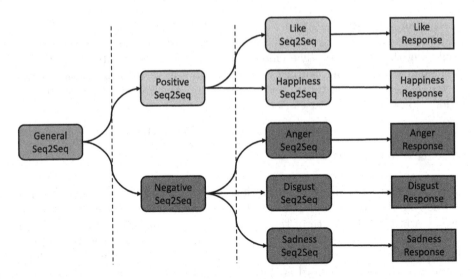

Fig. 2. Multi-Seq2Seq models with fine tune

under different emotion categories, we train five vanilla Seq2Seq models with attention mechanisms for five emotion categories respectively.

As shown in Fig. 2, all these models are pre-trained on the dataset of all post-response pairs at first, then fine-tuned on the dataset of pairs with the specific emotion. In more detail, before fine-tuning with five emotional datasets, we first fine-tune the general model to get two models with different emotional polarities. Then we fine-tune the positive model to get like and happy model, and fine-tune the negative model to get anger, disgust, and sad model.

Compared with like or happy datasets, the data of negative emotions such as anger or disgust has the smaller size, the emotions are expressed more subtly. For machine, it is easier to learn to generate negative responses than anger or disgust responses. Perhaps for this reason, we have achieved pretty good performance in negative response generation, especially for anger and disgust emotions.

Given a user-specific emotion, the corresponding model generates some candidates with this emotion. All candidates including those generated by other models will be fed into a ReRanker, which will be introduced later.

2.4 Rule-Based Model

The rule-based method is also called template-based method. As shown in Table 1, we construct many templates with blank for different emotion categories, where the blank is the object of emotions.

When we express emotions, our emotions usually point to some objects. So keywords and objects in posts are important to generate responses. We use the RUCNLP[2] tool to extract entities, and find keywords based on the dictionary

---

[2] http://183.174.228.47:8282/RUCNLP/.

Table 1. Examples of rule-based method

Post	**Hainan** tour is ruined [angry] [angry] [angry]
Like Response	I like **Hainan** most
Happy Response	I am very happy when I think of **Hainan**
Angry Response	I don't want to hear about **Hainan**, don't mention it!
Disgust Response	Super dislike **Hainan**!
Sad Response	**Hainan** broke my heart

and rules. These keywords and objects are combined with artificial templates to generate more fluent and point-explicit responses with emotions. If keywords are detected, these rule-based responses will be generated as the final submitted responses.

2.5   ReRanker

Now we have many candidates generated by models. We design a re-ranker to rank these candidates and select the response with the highest score as the final reply. As we mentioned before, emotion words are extremely important for explicit expression. We want our responses to satisfy fluency, coherence and emotional consistency. Now that the responses have ranked by generated probability, we only consider the emotional consistency and coherence. We use the combination of emotion score and coherence score as the metric to re-rank the response, and select the best response.

Emotion Score. We calculate the emotion score according to emotion dictionary. Based on the emotional vocabulary ontology library published by DUTIR [12], and emotion words extracted by χ^2 value from different emotion text data, we construct emotion vocabulary with the corresponding score for {*Like, Sad, Disgust, Anger, Happy*}. The scores reflect the importance of words in the specific emotion, which are composed of the weight given in the library and the frequency in training set. The explicit emotion words have higher scores than implicit emotion words. For example, "happy" has a greater weight than "joy" in happiness emotional dictionary. The emotion score of a sentence is the sum of the emotion words' scores.

In addition, we also consider the degree words to adjust emotion scores and categorize them into different levels. The degree words, such as "very", "a little", "not", can increase, decrease or reverse the emotional expression. The degree levels are reduced in the order of most, very, especially, little, inverse and others and the weight are set to 2, 1.5, 1.25, 0.5, −1, 1, respectively. Therefore given a sentence, the emotion score is calculated by:

$$\epsilon_m = \prod_{j \in index(m-1,m)} l_{y_j} \cdot \gamma_m \cdot w_m, \tag{12}$$

$$\mathcal{E}(\mathbf{y}) = \sum_{i=1}^{M} \epsilon_m, \tag{13}$$

where M are emotion words in the candidate response $\mathbf{y}$. $\mathcal{E}(\mathbf{y})$ and ϵ_m are the emotion scores of the candidates $\mathbf{y}$ and emotion word m, respectively. $index(m-1, m)$ is the index scope from the previous emotion word to the current emotion word. $index(0) = 0$ is for the first word. l_{y_j} is the level of degree word y_j in this scope to reflect the influence of increasing, decreasing or reversing the original emotion. w_m is the weight of emotion word m. γ_m indicates whether the emotion word m is in its corresponding emotional category. If the word m is in the corresponding dictionary of emotion (e.g., "happy" for happiness emotion), then we set γ_m as 1 to reflect this positive effect. Otherwise, we set γ_m as -1 to reflect the negative effect (e.g., "sad" occurs in happiness emotion).

Coherence Score. However, only the emotion score can not measure the quality of response comprehensively. For example, given "I won the prize." as the post, "I am so happy and excited." may get a higher emotion score, but "I am very happy that you won the prize." is more appropriate with coherent information than the former response. Therefore, we calculate the term similarity between the response and post as the coherence score, to encourage our model to generate results with consistent information. We select the number of same terms between the response and the post as the measure of consistency.

$$\mathcal{T}(\mathbf{y}) = Count(\mathbf{x}, \mathbf{y}), \tag{14}$$

where $Count(\cdot)$ counts the same term between post $\mathbf{x}$ and candidate response $\mathbf{y}$. Finally, the ranking score of $\mathbf{y}$ is computed by:

$$\Phi(\mathbf{y}) = \lambda \mathcal{E}(\mathbf{y}) + (1 - \lambda)\mathcal{T}(\mathbf{y}), \tag{15}$$

where the λ is set to 0.2 after many tests verified.

3   Experiment and Analysis

3.1   Implementation and Submissions

We submit 2 runs in this task. The settings of each run are shown as follows. Even we do not conduct adequate ablation experiments because of the limitation of the number of submitted runs, we can also learn the importance of every module by the comparison of two runs.

Table 2. Official CECG subtask results of the overall score and average score.

Team name	Label 0	Label 1	Label 2	Total	Overall score	Average score
1191_1	581	320	99	1,000	518	0.518
1191_2	831	109	60	1,000	229	0.229
AINTPU_1	716	200	84	1,000	367	0.336
CKIP_1	845	29	126	1,000	281	0.281
CKIP_2	840	28	132	1,000	292	0.292
IMTKU_1	580	248	172	1,000	592	0.592
IMTKU_2	954	32	14	1,000	60	0.060
TMUNLP_1	777	126	97	1,000	320	0.320
TUA1_1	443	293	264	1,000	821	0.821
TUA1_2	454	278	268	1,000	814	0.814
WUST_1	601	211	188	1,000	587	0.587
WUST_2	999	0	1	1,000	2	0.002
TKUIM_2	507	260	233	1,000	726	0.726
RUCIR_1	392	263	**345**	1,000	**953**	**0.953**
RUCIR_2	460	**342**	198	1,000	738	0.738

Table 3. Top 3 runs of official emotion-specific results on each emotion.

Emotion category	Team name	Label 0	Label 1	Label 2	Total	Overall score	Average score
Like	RUCIR_1	88	36	76	200	**188**	**0.940**
	RUCIR_2	96	44	60	200	164	0.820
	TKUIM_2	90	56	54	200	164	0.820
Sad	RUCIR_1	72	48	80	200	**208**	**1.040**
	TUA1_1	84	31	85	200	201	1.005
	RUCIR_2	83	57	60	200	177	0.885
Disgust	RUCIR_1	71	76	53	200	**182**	**0.910**
	TUA1_2	92	82	26	200	134	0.670
	TUA1_1	82	105	13	200	131	0.655
Anger	RUCIR_1	88	63	49	200	**161**	**0.805**
	TKUIM_2	112	45	43	200	131	0.655
	TUA1_2	85	107	8	200	123	0.615
Happy	TUA1_2	76	25	99	200	**223**	**1.115**
	TUA1_1	71	36	93	200	222	1.110
	RUCIR_1	73	40	87	200	214	1.070

- RUCIR_1: a combination of candidates from emotion-aware Seq2Seq model, multi-Seq2Seq model and rule-based model introduced in Sects. 2.2, 2.3 and 2.4 respectively, then reranked by ReRanker to get the top one.
- RUCIR_2: the top candidate of emotion-aware Seq2Seq introduced in Sect. 2.2. This is submitted as a baseline for RUCIR_1.

The released dataset contains 1,719,207 Weibo post-response pairs. After data pre-processing, there are 1,603,167 pairs in our dataset. We randomly select 5,000 pairs as the validation set and testing set respectively. The rest pairs compose training set. We construct two separate vocabularies for posts and responses by using 10,000 most frequent words on each side, covering 95.98% and 96.38% usage of words for posts and responses respectively. And the emotion vocabulary size is 500 in total for all emotions. The words out of vocabulary are replaced with a special token "<UNK>".

We use Tensorflow[3] to implement all models. A four-layered GRU cell with 1,024 dimensions is employed for both the encoder and the decoder. The dropout probability is set to 0.3. All model parameters are initialized with uniform distribution in $[-0.08, 0.08]$. Word embeddings and emotion embeddings are randomly initialized and learned during training with 200 dimensions and 50 dimensions respectively. All candidates are generated using beam search with 10 beam width. We train the models on NVIDIA TITAN Xp GPU using the Adam optimizer with an initial learning rate 5e−4 and a decay factor 0.9. The batch size is 64.

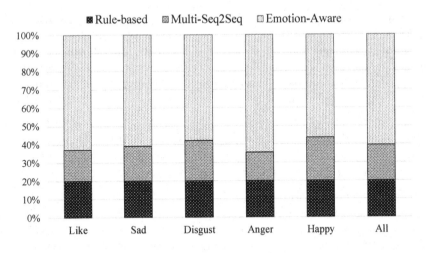

Fig. 3. The Proportion of Responses from Three Models in RUCIR_1

---

[3] https://www.tensorflow.org.

3.2   Results and Analysis

In the NTCIR-14 STC-3 CECG subtask [13], the submitted post-response pairs are evaluated by human annotation. The evaluation metrics are Fluency, Coherence and Emotion Consistency. The evaluation set has 200 posts and we submit the responses in five emotion categories except *other*.

Table 2 shows the overall results of all runs in CECG. Table 3 shows the top 3 runs of emotion-specific results on each emotion category. We can see that RUCIR_1 achieves best performances in overall results and in four of the five emotion-specific results. Moreover, the performance under the *happy* emotion category is also very close to the top.

As shown in Fig. 3, every model has contributed to the RUCIR_1 results. The contribution of emotion-aware Seq2Seq model is the most (about 60%), followed by Rule-based model (20%) and Multi-Seq2Seq model (about 20%). Note that we select the rule-based response as the final submitted response if there are keywords in the given post, thus the proportions of rule-based results are equal. Through our hybrid framework, we have taken advantage of both template-based method and data-driven methods, and achieved the best performance.

Even without the rule-based method and multi-Seq2Seq model, our emotion-aware Seq2Seq with emotion information still achieves fourth in all runs. And RUCIR_2 has more label 1 terms than others which means the responses generated by RUCIR_2 are more coherent and fluent. These prove the effectiveness of our model. We can infer that our improved seq2seq model can guarantee the fluency and coherence of response at least.

In addition, as shown in Table 3, our model has achieved pretty good performance in negative response generation, especially for anger and disgust emotions. As mentioned before, the quality of negative dataset is unsatisfactory. Thus, the pure data-driven approach does not work well. Compared with emotion-aware Seq2Seq model (RUCIR_2), the rule-based method and multi-seq2seq with fine tune (RUCIR_1) have more advantages in such a situation.

4   Conclusion

In this paper, we introduce our approaches in the CECG subtask of NTCIR-14 STC-3 task. We design a hybrid framework that includes three different models to generate responses and one re-rank to select the best response.

We introduce an emotion-aware Seq2Seq model with emotion factors and emotion words to generate responses. And we use the emotion score and coherence score as an additional feature to re-rank the response candidates. The experimental results verify the effectiveness of our methods.

In more detail, we use both the rule-based method and data-driven models. The rule-based method is the template-based method. In data-driven models, we design the multi-Seq2Seq model with fine-tune and the emotion-aware Seq2Seq model. We introduce the emotion-aware Seq2Seq model with emotion factors and emotion words to generate responses. And we use the emotion score and coherence score as an addition feature to re-rank the response candidates. The experimental results verify the effectiveness of our methods.

In the future, we will focus on several aspects: extracting other types of information from sentences, designing the more reasonable re-rank method, building a more advanced model to combine keywords extraction and keywords placement during the training.

Acknowledgements. Zhicheng Dou is the corresponding author. This work was supported by National Key R&D Program of China No. 2018YFC0830703, National Natural Science Foundation of China No. 61872370, and the Fundamental Research Funds for the Central Universities, and the Research Funds of Renmin University of China No. 2112018391.

References

1. Bahdanau, D., Cho, K., Bengio, Y.: Neural machine translation by jointly learning to align and translate. arXiv preprint arXiv:1409.0473 (2014)
2. Cho, K., et al.: Learning phrase representations using RNN encoder-decoder for statistical machine translation. arXiv preprint arXiv:1406.1078 (2014)
3. Hochreiter, S., Schmidhuber, J.: Long short-term memory. Neural Comput. **9**(8), 1735–1780 (1997)
4. Li, J., Galley, M., Brockett, C., Gao, J., Dolan, B.: A diversity-promoting objective function for neural conversation models. arXiv preprint arXiv:1510.03055 (2015)
5. Luong, M.T., Pham, H., Manning, C.D.: Effective approaches to attention-based neural machine translation. arXiv preprint arXiv:1508.04025 (2015)
6. Shang, L., Lu, Z., Li, H.: Neural responding machine for short-text conversation. arXiv preprint arXiv:1503.02364 (2015)
7. Shang, L., et al.: Overview of the NTCIR-13 short text conversation task (2017)
8. Shang, L., Sakai, T., Lu, Z., Li, H., Higashinaka, R., Miyao, Y.: Overview of the NTCIR-12 short text conversation task, pp. 473–484 (2016)
9. Song, Y., Li, C.T., Nie, J.Y., Zhang, M., Zhao, D., Yan, R.: An ensemble of retrieval-based and generation-based human-computer conversation systems. In: Proceedings of the Twenty-Seventh International Joint Conference on Artificial Intelligence, IJCAI-18, pp. 4382–4388. International Joint Conferences on Artificial Intelligence Organization, July 2018
10. Sutskever, I., Vinyals, O., Le, Q.V.: Sequence to sequence learning with neural networks. In: Ghahramani, Z., Welling, M., Cortes, C., Lawrence, N.D., Weinberger, K.Q. (eds.) Advances in Neural Information Processing Systems 27, pp. 3104–3112. Curran Associates, Inc. (2014)
11. Xing, C., et al.: Topic aware neural response generation. In: Thirty-First AAAI Conference on Artificial Intelligence (2017)
12. Xu, L., Lin, H., Pan, Y., Ren, H., Chen, J.: Constructing the affective Lexicon ontology. J. China Soc. Sci. Tech. Inf. **27**(2), 180–185 (2008)
13. Zhang, Y., Huang, M.: Overview of NTCIR-14 short text generation subtask: emotion generation challenge. In: Proceedings of the 14th NTCIR Conference (2019)
14. Zhou, H., Huang, M., Zhang, T., Zhu, X., Liu, B.: Emotional chatting machine: emotional conversation generation with internal and external memory. In: Thirty-Second AAAI Conference on Artificial Intelligence (2018)

We Want Web

THUIR at the NTCIR-14 WWW-2 Task

Yukun Zheng, Zhumin Chu, Xiangsheng Li, Jiaxin Mao, Yiqun Liu[✉],
Min Zhang, and Shaoping Ma

Department of Computer Science and Technology, Institute for Artificial Intelligence,
Beijing National Research Center for Information Science and Technology,
Tsinghua University, Beijing 100084, China
yiqunliu@tsinghua.edu.cn

Abstract. The THUIR team participated in both Chinese and English
subtasks of the NTCIR-14 We Want Web-2 (WWW-2) task. This paper
describes our approaches and results in the WWW-2 task. In the Chinese subtask, we designed and trained two neural ranking models on
the Sogou-QCL dataset. In the English subtask, we adopted learning to
rank models by training them on MQ2007 and MQ2008 datasets. Our
methods achieved the best performances in both Chinese and English
subtasks. Through further analysis of results, we find that our neural
models can achieve better performances in all navigational, informational
and transactional queries in Chinese subtask. In the English subtask, the
learning-to-rank methods have stronger modeling capabilities than BM25
by learning from effective hand-crafted features.

Keywords: Web search · Ad-hoc retrieval · Document ranking

1 Introduction

A lot of learning to rank approaches have been proposed to address document
ranking problem, such as AdaRank [41], LambdaMART [38] and etc. All these
learning to rank algorithms usually need to be trained with effective hand-crafted
features in the learning process. IR community has applied deep learning methods to advance state-of-the-art retrieval technologies. Guo et al. [13] suggested
that most of the recent neural ranking models can be generally classified into
two categories according to the network architectures: (1) *Representation-focused
model.* Models in this category first learn vector representations for textual
queries and candidate documents separately with deep neural networks. Then the
relevance is calculated by measuring the similarities between the two representations. This line of research includes DSSM [17], C-DSSM [34] and ARC-I [16],

This work is supported by the National Key Research and Development Program
of China (2018YFC0831700) and Natural Science Foundation of China (Grant No.
61622208, 61732008, 61532011).

M. P. Kato et al. (Eds.): NTCIR 2019, LNCS 11966, pp. 165–179, 2019.
https://doi.org/10.1007/978-3-030-36805-0_13

etc. (2) *Interaction-focused model.* ARC-II [16], DRMM [13], MatchPyramid [29] and K-NRM [40] belong to this category. The term-level interactions between queries and candidate documents are calculated first in these models. Then, the neural networks learn query-document matching patterns from these interactions. Mitra et al. [26] proposed to take advantage of both architectures in the Duet model. Luo et al. [23] showed the effectiveness of neural ranking models trained on large-scale weakly supervised data in ad-hoc retrieval. Self-attention mechanism [37] has been introduced into a number of NLP tasks, which helps models achieve better performances.

In the Chinese subtask, with the success of the neural methods in the ad-hoc retrieval task, we design a Deep Matching Model with Self-Attention (DMSA), which combines both the interaction-focused and representation-focused frameworks and incorporates both weakly supervised relevance and human relevance in the training process. Besides, we also design a Simple Deep Matching Model (SDMM) which sequentially models the interaction of the query and each sentence. Specifically, we apply a local matching layer to capture the exact matching and semantic matching signals. We applied these two models re-ranking on the top results of the baselines runs. Experiment results show SDMM's state-of-the-art performance among all the submitted runs [24].

In the English subtask, we try several learning to rank methods and BM25 because of the lack of large English datasets with relevance judgments. We submitted baseline run and another BM25 run based on a fine-grained document index as well as three runs of different learning-to-rank models. The submitted runs of learning to rank models, i.e., AdaRank [41], LambdaMART [38] and Coordinate Ascent [36], belong to pair-wise or list-wise methods, which are popular methods to be used in document ranking task. In our experiment, the results show that the learning to rank models perform much better than BM25 [24].

We publicly release the dataset we used in the WWW-2 task, the codes of our model implementations and ranking results[1] to researchers.

2   Related Work

2.1   Learning to Rank

After the traditional ranking models (e.g. BM25 [33], Language Model [43]), many IR researchers have focused on the learning-to-rank models [5,20]. Using the ideas of machine learning technologies, learning-to-rank models combine multiple ranking signals and become more effective and explainable. In general, learning-to-rank models can be categorized into three types: pointwise, pairwise and listwise. In the pointwise approaches which are popular in the earliest learning-to-rank models, each pair of query and document are treated as an independent individual. The pairwise approaches construct the optimized loss function using the relative relevance between two documents. Generally, the goal of learning is to minimize the number of miss-ordered document pairs. The

---

[1] https://github.com/zhengyk11/WWW2_THUIR_Runs.

third type of approaches, listwise approaches, consider all the documents associated with the same query as a unified entirety to learn and predict their ground truth labels. There are two types of listwise approaches according to the loss function. For the first type, the loss function is explicitly related to a particular evaluation metric (e.g. MAP [1], NDCG [18]). Since our commonly used metrics are discontinuous and non-differential, the common loss is the upper bound or approximation of a specific evaluation metric. SoftRank [35], SmoothRank [6] and Approximate Rank [32] are some typical instances of this type. For the second type, the loss function is not explicitly related to a particular evaluation metric. The loss reflects the inconsistency between the predicted ranking list and the ground truth ranking list instead. Typical examples include ListNet [4], ListMLE [39] and LambdaMART [3].

2.2   Neural Ranking Model

Compared to traditional ranking models (e.g., BM25 [33], SDM [25]), neural retrieval models are capable of automatically learning features from raw text and capturing more semantically relevant signals [19]. Existing retrieval models can generally be divided into two categories, namely representation-based models and interaction-based models [15]. Representation-based models aim to learn an integrated semantic representation of query and document. For example, Huang et al. [17] proposed a Deep Structured Semantic Model (DSSM), which is the first successful representation-based model for information retrieval. Its idea is to represent two input texts by using a multi-layer perceptron (MLP) transformation. Later on, Hu et al. [16] and Palangi et al. [27] proposed to use convolutional neural networks (CNN) and recurrent neural network to replace MLP, achieving relatively better ranking performance. Although representation-based models achieve fairly good ranking performance, they lose a lot of fine-grained semantic information (e.g., passage or sentence-level relevance [19]) due to the feature extraction architecture. To solve this problem, interaction-based models are proposed. They model the local interaction between query and document, which is able to capture more fine-grained semantic information. Specifically, the interaction function can be parametric or non-parametric [15], depending on whether it contains trainable parameters. A classic interaction-based model is DRMM [14], it applied matching histogram mapping to model the semantic relevance of the document to each query term and then aggregate them together into a relevance score. Matchpyramid [29] built a word-to-word similarity matrix and applied convolution neural networks to model the interaction. To model multi-level semantic matching between query and document, Xiong et al. [40] proposed KNRM by using kernel pooling strategies. It is further extent to EDRM [21] and Conv-KNRM [9] with knowledge graph and n-grams information, respectively. DeepRank [30] selectively considers the semantic matching occurring at query centric context. It can be considered as a neural-form BM25 model to some extent. HiNT [11] models passage-level information and accumulates to final relevance with an LSTM model, which is the first fine-grained passage modeling approach.

Table 1. Statistics of the dataset in our experiments. *Click* means the click relevance label from click model while *Manual* means human annotated label.

Dataset	QCL-Train	QCL-Test	NTCIR-13
#Query	534,655	2,000	100
#Doc	7,682,872	50,150	9,985
#Doc per query	14.37	25.08	99.85
Vocabulary size	821,768	445,885	211,957
Label type	Click	Click	Manual

3   Chinese Subtask

3.1   Dataset

To evaluate the performance of different retrieval models, we conduct experiments on a large-scale public available benchmark data (QCL) [44] and use the NTCIR-13 test set as the validation set in the Chinese subtask. Table 1 shows the statistics of the datasets. QCL is sampled from the query log of a popular Chinese commercial search engine Sogou. It contains weak relevance labels derived by five different click models for over 12 million query-document pairs. The number of query and document is about 500 thousand queries and more than 9 million documents. Prior works [8,40] have shown that such weak relevance labels derived by click models can be used to train and evaluate retrieval models. Thus, in our work, we utilize click relevance label to train our model. The click relevance labels of QCL are derived from five click models, TACM, PSCM, UBM, DBN, and TCM respectively. We use relevance inferred by PSCM to train the retrieval models because the PSCM has the best relevance estimation performance among these five alternatives. Besides, Sogou-QCL provides a smaller dataset with 2,000 queries and about 50 thousand documents, where all the query-document pairs have 4-point scaled relevance labels from human annotators. Similar to the evaluation settings used in [40], we utilize two different click relevance labels to evaluate our model on the test set of QCL. During the validation process, we uses click relevance labels from the same PSCM to evaluate our model.

3.2   Deep Matching Model with Self Attention

In the Chinese subtask, we design a deep matching model with self-attention mechanism (DMSA). Figure 1 shows the framework of DMSA, which consists of two weakly supervised relevance predictors, BM25 score predictor (BM25 predictor) and click model-based relevance predictor (CM predictor), and a multi-relevance fusion predictor. BM25 predictor and CM predictor are used to predict the BM25 score and click model-based relevance respectively and share the same framework as shown in Fig. 2. Multi-relevance fusion predictor are adopted to predict the human relevance based on the real BM25 score, the predicted BM25 score and the predicted score of click model-based relevance.

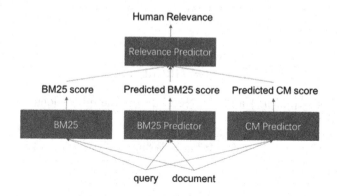

Fig. 1. The framework of DMSA.

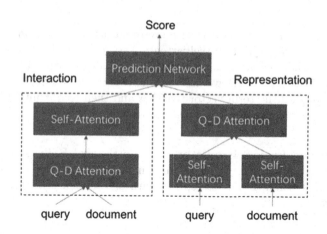

Fig. 2. The framework of weakly supervised relevance predictor.

Weak Supervised Predictor. In the weakly supervised predictor, we use an interaction-focused sub-model and a representation-focused sub-model to process the input of question and document terms simultaneously. Through each sub-model, we get two learned representations of the document. Then we use a multilayer perceptron to predict the weak relevance label based on the concatenation of the two learned representations of documents.

In the interaction-focused and representation-focused sub-models, we adopt the self-attention mechanism, which is very popular in NLP tasks, such as machine reading comprehension. We formulate the implementation of the attention mechanism in DMSA. Given a query $Q = \{q_1, ..., q_n\}$ and a document $D = \{d_1, ..., d_m\}$ as the inputs, where the query and the document consist of several terms, we first utilize a Gate Recurrent Unit (GRU) [7] to learn the context-aware representations of the texts.

$$u_1, ..., u_n = \text{GRU}(q_1, ..., q_n) \tag{1}$$

$$v_1, ..., v_m = \text{GRU}(d_1, ..., d_m) \tag{2}$$

Given $U = \{u_1, ..., u_n\}$ and $V = \{v_1, ..., v_m\}$, the query-document attention is conducted as follows:

$$s_j^i = W^q u_j \odot W^d v_i \tag{3}$$

$$a_j^i = \exp(s_j^i) / \sum_{t=1}^{n} \exp(s_i^t) \tag{4}$$

$$c_j = \sum_{i=1}^{n} a_j^i u_i \tag{5}$$

$$h_j = W^h [c_j, v_j] \tag{6}$$

where $H = \{h_1, ..., h_m\}$ is the learned representation of the document after the query-document attention. In the self attention stage, we feed the term sequence of the query or the document as the input to conduct the attention with itself.

In the prediction network, we first get the representation vector of the document by adding all the term vector together and then feed it into a multilayer perceptron to predict the weak relevance label.

Multi-relevance Fusion Predictor. We use the real BM25 score, the predicted BM25 score and the predicted click model-based relevance score as the input and adopt a multilayer perceptron with one hidden layer to predict the human relevance.

3.3 Simple Deep Matching Model

Figure 3 shows the framework of our simple deep matching model (SDMM), which contains a local matching layer and a recurrent neural network (RNN) layer. The local matching layer aims to capture the semantic matching between query and sentence. The basic idea is to follow IR heuristics [12, 28] and qualify them into a semantic representation. Sentence representations are then aggregated sequentially by using a recurrent neural network for estimating relevance score. The aggregation layer is fed with the semantic representation and position embedding of each sentence.

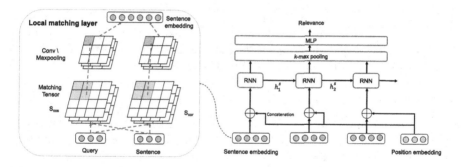

Fig. 3. The framework of SDMM.

Local Matching Layer: Following the idea in [11], we apply term-level inter-action matrix with both exact query matching and semantic query matching. Specifically, for a given query $\mathbf{q} = [w_1, w_2, ..., w_m]$ and a document $\mathbf{d}$ with T sentences, where each sentence is $\mathbf{s} = [v_1, v_2, ..., v_n]$, we construct a semantic matching matrix M^{cos} and an exact matching matrix M^{xor}, which are defined as follows:

$$M_{ij}^{cos} = cos(w_i, v_j), \tag{7}$$

$$M_{ij}^{xor} = \begin{cases} 1, & w_i = v_j \\ 0, & otherwise \end{cases} \tag{8}$$

Exact matching and semantic matching provide critical signals for informa-tion retrieval as suggested by [12,28]. To further incorporate term importance to the input, we extend each element M_{ij} to a three-dimensional representation vector $S_{ij} = [x_i, y_j, M_{ij}]$ by concatenating two term embeddings as in [11], where $x_i = w_i * \mathbf{W}_c$ and $y_j = v_j * \mathbf{W}_c$. $\mathbf{W}_c$ is a compressed matrix to be learned during training. The proximity of each word matching is retained in these matching matrices.

Based on two interaction matrices, we further apply CNN to generate local relevance embedding, which is also called sentence embedding. Note that CNN is more efficient than spatial GRU applied in [11] and it can also capture the relation between several adjacent words. The final sentence embedding is repre-sented by concatenating the signals from two interaction matrices:

$$\mathbf{s} = [CNN(\mathbf{S}^{cos}), CNN(\mathbf{S}^{xor})] \tag{9}$$

Aggregation Layer: We then concatenate the sentence embedding from local matching layer with position embedding $\mathbf{p}$. Position embedding can be looked up based on the serial number of the input sentence. More specifically, we divide the sequences into a fixed number of blocks and sentences in the same block are assigned with the same serial number for looking up the position embedding. Our model sequentially processes each input sentence by transferring the concatenated embedding $g_t = [s_t, p_t]$ into a RNN module:

$$h_t^s = RNN(h_{t-1}^s, g_t), t = 1, ..., T \tag{10}$$

Where T is the number of total sentences of a document. Modeling sentences by RNN module is able to capture the context information in neighboring sentences. The RNN we used is GRU.

The hidden state $h_{1:T}^s$ are then utilized to estimate relevance by a k-max pooling layer and a full connected layer. k-max pooling layer selects top-k signals over all the sentences and full connected layer maps hidden states to a relevance score.

3.4   Experiment Setup

DMSA. We train the DMSA model in a point-wise and multi-task method to simultaneously predict human relevance, BM25 score and click model-based relevance of a query-document pair. We adopt mean squared error (MSE) as the loss function with adadelta [42] as the optimizer. The learning rate is 0.01 and the dropout rate is 0.2. The dimension of the embedding layer is 200 with initial word embeddings pre-trained on Sogou-QCL using word2vec, and the hidden size is 100. The batch size is 20 during model training.

SDMM. We train the SDMM model in a point-wise learning method with mean squared error (MSE) as the loss function. The parameters are optimized by adadelta, with a batch size of 80 and a learning rate of 0.1. The dimension of the embedding layer is 50 and it is initialized with the word2vec trained on a Chinese Wikipedia dataset[2]. The dimension of other the hidden vectors is 128. The convolution neural network applies filters with window sizes from 2 to 5, where each filter has 64 feature maps. The recurrent neural network we used is Gated Recurrent Unit (GRU). In addition, the number of candidate documents of each query in NTCIR dataset is large, so we first retrieve top 40 documents with highest BM25 scores and then rerank them based on our model.

We implement both DMSA and SDMM by PyTorch[3]. Early stopping with a patience of 10 epochs is adopted during the training processes of two models.

3.5   Submitted Runs and Evaluation

We submitted 5 runs which were tested by the DMSA and SDMM models based on different numbers of top results in the baseline run, as shown in Table 2.

---

[2] http://download.wikipedia.com/zhwiki.

[3] https://github.com/pytorch/pytorch.

Table 2. Overview of runs in the Chinese subtasks.

Run	Model	Re-rank range
THUIR-C-CO-MAN-Base-1	DMSA	10
THUIR-C-CO-MAN-Base-2	DMSA	100
THUIR-C-CO-MAN-Base-3	DMSA	45
THUIR-C-CO-CU-Base-4	SDMM	100
THUIR-C-CO-CU-Base-5	SDMM	40

Table 3. Evaluation of runs in the Chinese subtasks. The table shows the mean value and the rank of the metric among all 10 runs submitted in the subtask. */** indicates that the improvement is statistically significant at $p < 0.05$ or $p < 0.01$ level using the two-tailed and pairwise t-test.

Run	nDCG@10		Q@10		nERR@10	
THUIR-C-CO-CU-Base-5	**0.4916****	1	**0.4610****	1	**0.6374****	1
THUIR-C-CO-MAN-Base-2	0.4835**	3	0.4604**	2	0.5973*	4
THUIR-C-CO-MAN-Base-1	0.4748**	4	0.4479**	4	0.6019**	3
THUIR-C-CO-MAN-Base-3	0.4706**	5	0.4364**	5	0.5829*	5
THUIR-C-CO-CU-Base-4	0.4458*	9	0.4189**	9	0.5663	7
Official BM25 baseline	0.3545	-	0.3080	-	0.4869	-

Table 4. Examples of three query categories in the Chinese subtask.

qid	Query	Caterogy
4	小米官网 (Xiaomi official website)	Navigational
11	科目三通过率下降 (Pass rate of driver license test three declines)	Informational
1	万圣节图片 (Halloween picture)	Transactional

Table 5. Model performances under three query categories in MSnDCG@10. The improvements are calculated compared to official BM25 baseline. */** indicates that the improvement is statistically significant at $p < 0.05$ or $p < 0.01$ level using the two-tailed and pairwise t-test.

Run	Navigational		Informational		Transactional	
THUIR-C-CO-MAN-Base-1	0.3877	23.6%	0.5082**	29.8%	0.4601**	39.8%
THUIR-C-CO-MAN-Base-2	**0.4274**	36.2%	**0.5354**	36.8%	0.4478*	36.0%
THUIR-C-CO-MAN-Base-3	0.3194	1.8%	0.5141**	31.3%	0.4575**	39.0%
THUIR-C-CO-CU-Base-4	0.3610	15.1%	0.5096*	30.2%	0.4044	22.8%
THUIR-C-CO-CU-Base-5	0.4000	27.5%	0.5233**	33.7%	**0.4791****	45.6%
Official BM25 baseline	0.3137	-	0.3915	-	0.3291	-

Table 3 shows the evaluation results and ranks of our five submitted runs in the Chinese subtask. *THUIR-C-CO-CU-Base-5* achieves the best performance among all submitted runs, which is generated by SDMM model trained on weakly supervised data.

We manually classified the valid 75 queries in the Chinese subtask into three categories according to [2]: *navigational, informational* and *transactional.* There are 6 navigational queries, 32 informational and 37 transactional ones. Table 4 shows three examples of these query categories in the Chinese subtask. Table 5 shows the performances of our submitted runs and the official BM25 baseline run under different query categories. Since the results are similar among the three metrics, we only report the performances in MSnDCG@10. We can see that DMSA in *THUIR-C-CO-MAN-Base-2* achieves the best ranking performance in both navigational and informational queries, while SDMM in *THUIR-C-CO-CU-Base-5* outperforms models of others runs under the transactional queries. All five runs we submitted get better performances than baseline in three query categories.

4   English Subtask

In English subtask, we adopted learning-to-rank model. Figure 4 shows its framework. We introduce the details of our models in this section.

Table 6. The features extracted for training learning-to-rank models

ID	Features
1	TF (Term frequency)
2	IDF (Inverse document frequency)
3	TF * IDF
4	DL (Document length)
5	BM25
6	LMIR.ABS
7	LMIR.DIR
8	LMIR.JM

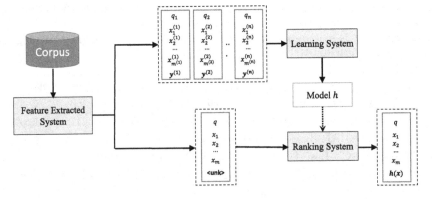

Fig. 4. The framework of learning-to-rank models.

4.1   Features Extraction

First, we preprocessed the queries and the html files of documents by lowercasing, tokenization, removing stop words, and stemming. These preprocessing methods are all implemented with the NLTK toolkit[4]. Next, we merged synonymous terms to make the extracted features more reliable and useful. To train learning-to-rank model, we extracted features listed in Table 6. Besides TF/IDF, document length and BM25, we also extracted three kinds of language model based on the implementation of Zhai et al. [43]. In another word, we extracted eight-dimensional features for four fields of a document: the whole document content, anchor text, title, and URL. Finally, we obtained $4 \times 8 = 32$ features for each document in total.

Table 7. The statistics of the dataset used in the training, validation and test processes

Dataset	Training set	Validation set	Test set
#Query	2,476	100	80
#Query-doc pair	55,954	20,980	78,673
Average #doc per query	22.6	209.8	983.4
Max #doc per query	100	265	999
Min #doc per query	1	137	929

4.2   Dataset

We chose the MQ2007 and MQ2008 [31] as our training set. Although they provide the features we required, we calculated these features with our own algorithms to ensure the consistency with the validation and test sets. At the same time, we used the NTCIR-13 WWW English testset [22] and its annotation results as our validation set, as its construction process is almost the same as

Table 8. Evaluation of runs in the English subtasks. The mean value and the rank of the metric are shown among all 19 runs submitted in the subtask. *LM* and *CA* means *LambdaMART* and *Coordinate Ascent* respectively. All the differences are not significant at $p < 0.05$ level using the pairwise and two-tailed t-test.

Run	Model	nDCG@10		Q@10		nERR@10	
THUIR-E-CO-MAN-Base-1	AdaRank	0.3444	4	0.3249	6	**0.5048**	1
THUIR-E-CO-MAN-Base-2	LM	0.3512	2	**0.3391**	1	0.5026	2
THUIR-E-CO-MAN-Base-3	CA	**0.3536**	1	0.3256	4	0.4805	4
THUIR-E-CO-PU-Base-4	BM25	0.3294	8	0.3161	8	0.4692	8
THUIR-E-CO-PU-Base-5	baseline	0.3258	11	0.3043	11	0.4779	5

---

[4] https://www.nltk.org/.

that of this year's testset. Table 7 shows the statistics of our training, validation and test datasets. We can notice that the average size of query-document pairs per query in training set is much smaller than that in test set. We consider that the size imbalance of datasets is less likely to harm the training effect of models because the evaluation of the ranking list usually only focuses on the topmost results rather than the whole list.

Table 9. The queries in which the difference value between the MSnDCG@10 of *THUIR-E-CO-MAN-Base-3* and that of baseline run is more than 0.3.

qid	Query	THUIR-E-CO-MAN-Base-3	Baseline	Difference
0051	Prednisone	0.6290	0.1181	0.5109
0060	Car-parts.com	0.6346	0.1639	0.4707
0003	Women's clothing winter	0.6650	0.2773	0.3877
0027	Is it okay to drink yogurt after eating persimmon?	0.4279	0.1035	0.3244
0021	International gold price	0.7339	0.4098	0.3241

4.3   Methods and Results

Ranklib [10] package is the toolkit we used to implement the learning-to-rank algorithms, which contains the fine-designed interfaces of many algorithms (e.g. LambdaMART, ListNet, AdaRank). We chose the LambdaMART, AdaRank, and Coordinate Ascent as the methods of our final submissions, because these models performed well on validation set. In the meantime, we submitted the baseline run and another BM25 run based on a fine-grained document index. As we know, the quality of background corpus and the preprocessing process are two important factors to affect the quality of BM25 index. To obtain high-quality BM25 index, in our fine-grained BM25 model, we built the background corpus with the randomly selected several millions html files in ClueWeb12 corpus, to confirm that the html files we interest and the those in background corpus are similar, while the preprocessing part was done with the methods introduced in Sect. 4.1. Table 8 shows the performance of our runs in the English subtask, including the mean metric values and the ranks among the all 19 runs submitted in the English subtask. It indicates that our three learning-to-rank methods achieve the best performances among all runs submitted in the English subtask, while there is no significant difference between them. At the same time, we can find that the LambdaMART method, which is known as the state-of-the-art learning-to-rank algorithm, always performs in very high levels regardless of the evaluation metric used.

4.4   Case Study

Here we list some examples to show the positive cases of the runs based on learning-to-rank methods compared to the traditional BM25 run (i.e., the official baseline run). Due to the space limitation, we focus on the *THUIR-E-CO-MAN-Base-3* run v.s. baseline run in the MSnDCG@10 measurement. Table 9 shows the representative cases where the *THUIR-E-CO-MAN-Base-3* run outperforms the baseline run. For the query "prednisone", some of the top-rank documents in the baseline run are spam pages. In these webpages, the HTML files include a lot of hidden links whose anchor texts contain the word "prednisone". The baseline run cannot handle this cheating behavior and treats them as high ranking documents. In comparison, the *THUIR-E-CO-MAN-Base-3* run successfully recognizes and gives them low ranking scores. The query with qid 0060 also shows that the learning-to-rank method outperforms the BM25. The query "car-parts.com" is a navigational query, so the most relevant result to the query is the corresponding website. The learning-to-rank methods depict the information of the URL field explicitly, so they get better ranking performance in the navigational cases than BM25.

5 Conclusion

In the NTCIR-14 WWW-2 task, we participated and got the best performances of runs in both Chinese and English subtasks. In the Chinese subtask, we designed two deep ranking models, which have been shown to be effective in ad-hoc retrieval. By further analysis, we find that our models can help improve the document ranking lists for all three query categories: *navigational, informational* and *transactional*. In the English subtask, we adopt learning to rank methods and trained them on MQ2007 and MQ2008 datasets. Through a case study, we find that the learning to rank methods can generate better ranking based on a number of valid hand-crafted features. In the future, we would like to investigate how to better combine human relevance labels and weakly supervised relevance labels in the ad-hoc retrieval task and how to better take fine-grained matching signals into our ranking models.

References

1. Baeza-Yates, R., Ribeiro-Neto, B., et al.: Modern Information Retrieval, vol. 463. ACM Press, New York (1999)
2. Broder, A.: A taxonomy of web search. In: ACM SIGIR Forum, vol. 36, pp. 3–10. ACM (2002)
3. Burges, C.J.: From RankNet to LambdaRank to LambdaMART: an overview. Learning **11**(23–581), 81 (2010)
4. Cao, Z., Qin, T., Liu, T.Y., Tsai, M.F., Li, H.: Learning to rank: from pairwise approach to listwise approach. In: ICML'07 (2007)
5. Chapelle, O., Chang, Y.: Yahoo! learning to rank challenge overview. In: Proceedings of the Learning to Rank Challenge, pp. 1–24 (2011)
6. Chapelle, O., Wu, M.: Gradient descent optimization of smoothed information retrieval metrics. Inf. Retrieval **13**(3), 216–235 (2010)
7. Cho, K., et al.: Learning phrase representations using RNN encoder-decoder for statistical machine translation. arXiv preprint arXiv:1406.1078 (2014)

8. Chuklin, A., Markov, I., Rijke, M.d.: Click models for web search. Synth. Lect. Inf. Concepts Retrieval Serv. **7**(3), 1–115 (2015)
9. Dai, Z., Xiong, C., Callan, J., Liu, Z.: Convolutional neural networks for soft-matching N-grams in ad-hoc search. In: Proceedings of the Eleventh ACM International Conference on Web Search and Data Mining, pp. 126–134. ACM (2018)
10. Dang, V.: The Lemur project-Wiki-RankLib. Lemur Project (2012)
11. Fan, Y., Guo, J., Lan, Y., Xu, J., Zhai, C., Cheng, X.: Modeling diverse relevance patterns in ad-hoc retrieval. In: International ACM SIGIR Conference on Research and development in Information Retrieval, pp. 375–384 (2018)
12. Fang, H., Tao, T., Zhai, C.: A formal study of information retrieval heuristics. In: Proceedings of the 27th Annual International ACM SIGIR Conference on Research and Development in Information Retrieval, pp. 49–56. ACM (2004)
13. Guo, J., Fan, Y., Ai, Q., Croft, W.B.: A deep relevance matching model for ad-hoc retrieval. In: CIKM'16 (2016)
14. Guo, J., Fan, Y., Ai, Q., Croft, W.B.: A deep relevance matching model for ad-hoc retrieval. In: ACM International on Conference on Information and Knowledge Management (2016)
15. Guo, J., et al.: A deep look into neural ranking models for information retrieval. arXiv preprint arXiv:1903.06902 (2019)
16. Hu, B., Lu, Z., Li, H., Chen, Q.: Convolutional neural network architectures for matching natural language sentences. In: NIPS'14 (2014)
17. Huang, P.S., He, X., Gao, J., Deng, L., Acero, A., Heck, L.: Learning deep structured semantic models for web search using clickthrough data. In: CIKM'13 (2013)
18. Järvelin, K., Kekäläinen, J.: Cumulated gain-based evaluation of IR techniques. ACM Trans. Inf. Syst. (TOIS) **20**(4), 422–446 (2002)
19. Li, X., Mao, J., Wang, C., Liu, Y., Zhang, M., Ma, S.: Teach machine how to read: reading behavior inspired relevance estimation. SIGIR (2019)
20. Liu, T.Y., et al.: Learning to rank for information retrieval. Found. Trends® Inf. Retrieval **3**(3), 225–331 (2009)
21. Liu, Z., Xiong, C., Sun, M., Liu, Z.: Entity-duet neural ranking: understanding the role of knowledge graph semantics in neural information retrieval. arXiv preprint arXiv:1805.07591 (2018)
22. Luo, C., Sakai, T., Liu, Y., Dou, Z., Xiong, C., Xu, J.: Overview of the NTCIR-13 we want web task. NTCIR-13 (2017)
23. Luo, C., Zheng, Y., Mao, J., Liu, Y., Zhang, M., Ma, S.: Training deep ranking model with weak relevance labels. In: Huang, Z., Xiao, X., Cao, X. (eds.) ADC 2017. LNCS, vol. 10538, pp. 205–216. Springer, Cham (2017). https://doi.org/10.1007/978-3-319-68155-9_16
24. Mao, J., Sakai, T., Luo, C., Xiao, P., Liu, Y., Dou, Z.: Overview of the NTCIR-14 we want web task. In: Proceedings of the 14th NTCIR Conference on Evaluation of Information Access Technologies (2019)
25. Metzler, D., Croft, W.B.: A Markov random field model for term dependencies. In: Proceedings of the 28th Annual International ACM SIGIR Conference on Research and Development in Information Retrieval, pp. 472–479. ACM (2005)
26. Mitra, B., Diaz, F., Craswell, N.: Learning to match using local and distributed representations of text for web search. In: WWW'17 (2017)
27. Palangi, H., et al.: Deep sentence embedding using long short-term memory networks: analysis and application to information retrieval. IEEE/ACM Trans. Audio Speech Lang. Process. (TASLP) **24**(4), 694–707 (2016)
28. Pang, L., Lan, Y., Guo, J., Xu, J., Cheng, X.: A deep investigation of deep IR models. arXiv preprint arXiv:1707.07700 (2017)

29. Pang, L., Lan, Y., Guo, J., Xu, J., Wan, S., Cheng, X.: Text matching as image recognition. In: AAAI'16 (2016)
30. Pang, L., Lan, Y., Guo, J., Xu, J., Xu, J., Cheng, X.: DeepRank: a new deep architecture for relevance ranking in information retrieval. In: CIKM'17 (2017)
31. Qin, T., Liu, T.Y.: Introducing LETOR 4.0 datasets. arXiv preprint arXiv:1306.2597 (2013)
32. Qin, T., Liu, T.Y., Li, H.: A general approximation framework for direct optimization of information retrieval measures. Inf. Retrieval 13(4), 375–397 (2010)
33. Salton, G., McGill, M.J.: Introduction to Modern Information Retrieval. McGraw-Hill, New York (1983)
34. Shen, Y., He, X., Gao, J., Deng, L., Mesnil, G.: Learning semantic representations using convolutional neural networks for web search. In: WWW'14 (2014)
35. Taylor, M., Guiver, J., Robertson, S., Minka, T.: SoftRank: optimizing non-smooth rank metrics. In: Proceedings of the 2008 International Conference on Web Search and Data Mining, pp. 77–86. ACM (2008)
36. Uysal, I., Croft, W.B.: User oriented tweet ranking: a filtering approach to microblogs. In: Proceedings of the 20th ACM International Conference on Information and Knowledge Management, pp. 2261–2264. ACM (2011)
37. Vaswani, A., et al.: Attention is all you need. In: NIPS'17, pp. 5998–6008 (2017)
38. Wu, Q., Burges, C.J., Svore, K.M., Gao, J.: Adapting boosting for information retrieval measures. Inf. Retrieval 13(3), 254–270 (2010)
39. Xia, F., Liu, T.Y., Wang, J., Zhang, W., Li, H.: Listwise approach to learning to rank: theory and algorithm. In: Proceedings of the 25th International Conference on Machine Learning, pp. 1192–1199. ACM (2008)
40. Xiong, C., Dai, Z., Callan, J., Liu, Z., Power, R.: End-to-end neural ad-hoc ranking with Kernel pooling. In: SIGIR'17 (2017)
41. Xu, J., Li, H.: AdaRank: a boosting algorithm for information retrieval. In: Proceedings of the 30th Annual International ACM SIGIR Conference on Research and Development in Information Retrieval, pp. 391–398. ACM (2007)
42. Zeiler, M.D.: ADADELTA: an adaptive learning rate method. arXiv preprint arXiv:1212.5701 (2012)
43. Zhai, C., Lafferty, J.: A study of smoothing methods for language models applied to ad hoc information retrieval. In: ACM SIGIR Forum, vol. 51, pp. 268–276. ACM (2017)
44. Zheng, Y., Fan, Z., Liu, Y., Luo, C., Zhang, M., Ma, S.: Sogou-QCL: a new dataset with click relevance label. In: SIGIR'18 (2018)

Fine-Grained Numeral Understanding in Financial Tweet

Final Report of the NTCIR-14 FinNum Task: Challenges and Current Status of Fine-Grained Numeral Understanding in Financial Social Media Data

Chung-Chi Chen[1]([⊠]) [ID], Hen-Hsen Huang[2,4] [ID], Hiroya Takamura[3] [ID], and Hsin-Hsi Chen[1,4] [ID]

[1] Department of Computer Science and Information Engineering,
National Taiwan University, Taipei, Taiwan
cjchen@nlg.csie.ntu.edu.tw, hhchen@ntu.edu.tw
[2] Department of Computer Science, National Chengchi University, Taipei, Taiwan
hhhuang@nccu.edu.tw
[3] Artificial Intelligence Research Center,
National Institute of Advanced Industrial Science and Technology, Tokyo, Japan
takamura.hiroya@aist.go.jp
[4] MOST Joint Research Center for AI Technology and All Vista Healthcare,
Taipei, Taiwan

Abstract. NTCIR-14 FinNum task aims to disambiguate the meanings of the numerals in financial social media data from both coarse-grained taxonomy and fine-grained taxonomy. This task attracted 13 teams from 15 institutions in 6 different countries to register, received 16 submissions from 9 participants, and finally accepted 6 papers from participants. In this paper, we provide the analysis of the proposed dataset in depth, and discuss the challenges of fine-grained numeral understanding in financial social media data.

Keywords: Numeral understanding · Financial social media · Numeral corpus

1 Introduction and Motivation

When analyzing a financial instrument, investors always focus on two aspects, fundamental and technical. Investors using fundamental analysis attempt to evaluate the intrinsic value of the financial instrument. For the security of company, they may focus on the numerals in financial statements. For the treasury bond, they may evaluate the price depending on US Fed Funds Target Rate. Those who use technical analysis may employ the technical indicator like moving average (MA), relative strength index (RSI), and so on. No matter which analysis method investors use, numeral plays an important role, and provides much pivotal information in financial data.

© Springer Nature Switzerland AG 2019
M. P. Kato et al. (Eds.): NTCIR 2019, LNCS 11966, pp. 183–192, 2019.
https://doi.org/10.1007/978-3-030-36805-0_14

Table 1. Statistics of FinNum 2.0

Category	Subcategory	Train	Dev.	Test	Total	Ratio
Monetary		2467	261	459	3187	35.94%
	Money	589	52	95	736	8.30%
	Quote	792	89	152	1033	11.65%
	Change	143	8	25	176	1.98%
	Buy price	319	36	60	415	4.68%
	Sell price	103	10	22	135	1.52%
	Forecast	270	33	52	355	4.00%
	Stop loss	25	4	6	35	0.39%
	Support or resistance	226	29	47	302	3.41%
Percentage		838	105	170	1113	12.55%
	Relative	585	70	112	767	8.65%
	Absolute	253	35	58	346	3.90%
Option		169	11	22	202	2.28%
	Exercise price	113	5	14	132	1.49%
	Maturity date	56	6	8	70	0.79%
Indicator		167	22	27	216	2.44%
Temporal		2364	253	401	3018	34.03%
	Date	2079	223	351	2653	29.92%
	Time	285	30	50	365	4.12%
Quantity		741	87	154	982	11.07%
Product/Version		114	14	22	150	1.69%
		6860	753	1255	8868	100.00%

Numeral contains much important information in financial domain. For example, investors may use price-earnings ratio (P/E ratio) to evaluate the value of security of certain company, where both P/E ratio and the value of security are numeric information. For the purpose of understanding the fine-grained numeric information in social media data, we adopt the taxonomy for numerals [5]. In this numeral taxonomy, we classify numerals into 7 categories and further extend several of these categories into subcategories. Especially, the most important category, Monetary, is extended into 8 subcategories. (T1) is an instance that contains several numerals, and the categories of the numerals are dissimilar. For example, 8 is the numeral about quantity, 17.99 is about stop loss price, 200 is the parameter of technical indicator, and 1 is the price of stock. In such a short sentence, there are 4 kinds of numerals. That shows the importance of numerals in financial narrative.

(T1) 8 breakouts: $CHMT (stop: $17.99), $FLO (200-day MA), $OMX (gap), $SIRO (gap). One sub-$1 stock. Modest selection on attempted swing low

Table 1 shows the target categories and subcategories. In the FinNum task, the position of a numeral in a tweet is given in advance. Participants are asked to disambiguate its category. This task is further separated into two subtasks defined as follows.

- **Subtask 1:** Classify a numeral into 7 categories, i.e., Monetary, Percentage, Option, Indicator, Temporal, Quantity and Product/Version Number.
- **Subtask 2:** Extend the classification task to the subcategory level, and classify numerals into 17 classes, including Indicator, Quantity, Product/Version Number, and all subcategories.

Micro-averaged F-score and macro-averaged F-scores are adopted for evaluating the classification performance of participants' runs.

The application scenarios of the proposed tasks were also demonstrated [7], including market price prediction [5] and the test of informativeness [2]. The results show that the usefulness of sorting out the fine-grained numeral information from financial social media data.

2   Corpus Creation

We collected the data from StockTwits[1], one of the social trading platform for investors to share their ideas and strategies. Two experts were involved in the annotating process. The dataset, FinNum 2.0[2], used in this shared task, contains only the numerals in full agreement. There are 4,072, 457, and 753 tweets in training set, development set, and test set, respectively. Note that, there are at least one cashtag, e.g., $AAPL is a cashtag stands for the stock of Apple Inc., and at least one numeral in each tweet. There are total 8,868 annotated numerals in FinNum 2.0. The statistics of FinNum 2.0 is shown in Table 1. The annotations are licensed under the Creative Commons Attribution-Non-Commercial-ShareAlike 4.0 International (CC BY-NC-SA 4.0) license, and provided for academic usage.

3   Methods and Analysis

In this section, we analyze some methods independently, and compare the performance of different methods. The experimental results of each model are shown in Tables 2 and 3.

3.1   Enriched CNN

Ait Azzi and Bouamor [1] got the first place in both subtasks 1 and 2 with their Enriched-CNN (E-CNN) model. They use a sequence labeling technique to solve

---

[1] https://stocktwits.com/.

[2] http://nlg.csie.ntu.edu.tw/nlpresource/FinNum/.

Table 2. Experimental results of subtask 1

Method	Micro F1 (%)	Macro F1 (%)
E-CNN [1]	93.94	90.05
Hybrid NN [15]	91.87	87.94
RNN with CNN Filter [10]	89.72	80.93
Attentive RNN [13]	86.45	78.09
SVM [14]	74.02	63.71
Unsupervised [12]	74.27	63.53
Word-based CNN [5]	55.90	51.67

Table 3. Experimental results of subtask 2

Method	Micro F1 (%)	Macro F1 (%)
Fusion model [1]	87.17	82.40
Hybrid NN [15]	83.03	77.90
Attentive RNN	80.24	74.11
RNN with CNN Filter [10]	79.12	72.51
SVM [14]	60.88	52.93
Unsupervised [12]	63.67	51.90
Char-based CNN [5]	43.75	31.12

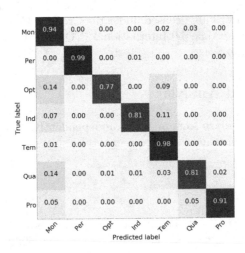

Fig. 1. Confusion matrix of the E-CNN model

the task, and concatenate pre-trained word embedding, character embedding extracted by CNN, POS tags, and hand-crafted features as the representation of a token. Figure 1 shows the confusion matrix of the E-CNN model in subtask 1. Only the accuracy of the Option category is lower than 80%. The results also show that the Percentage and Temporal are the easier categories in our taxonomy. The reason for these results may be that the numerals in Percentage category always follow by a symbol such as %, and the numerals in the Temporal category has some patterns like YYYY/MM/DD.

For subtask 2, the proposed fusion architecture takes the predictions of subtask 1 as the features for subtask 2. Figure 2 shows the confusion matrix of their model. Although the performances of several subcategories achieve over 80%, sorting out some meaningful subcategories is still a challenge for the model. For example, selling price and stop loss subcategories only get 50% of accuracy, and support or resistance price only get 70% of accuracy. The performance of the subcategories, exercise price and maturity date, in the Option category are only 71% and 62% of accuracy, respectively. The results show that although we can get good performance for many general (sub)categories, we still need to design the tailor-made solutions for the domain-oriented subcategories.

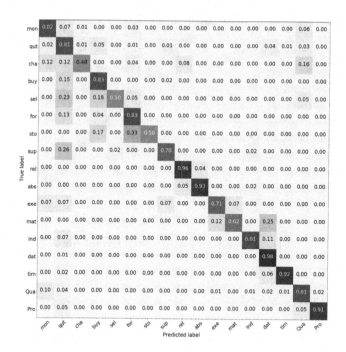

Fig. 2. Confusion matrix of the Fusion model

3.2 Hybrid NN

Wu et al. [15] present a Hybrid NN model with several features for representation, including normalization, orthography, format, pre-suffix, brown cluster, and recognizers-text type. Figure 3 shows the confusion matrix of their model. Comparing with the Fusion model, Hybrid NN model performs better in domain-specific subcategories and worse in general subcategories. In the five subcategories with the fine-grained opinions of individual investors [2], the performances of the Hybrid NN model are better than those of the Fusion model. Furthermore, the Hybrid model does not confuse the exercise price with the maturity date in the Option category. This is also different from the E-CNN model.

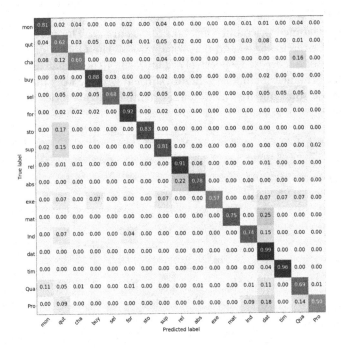

Fig. 3. Confusion matrix of the Hybrid NN

3.3 SVM

Wang et al. [14] use SVM mdoel to solve the task. Compare with Fusion model and Hybrid NN models, we find that the performance of SVM in Indicator category is the same as that of the Fusion model, and is better than that of the Hybrid NN model.

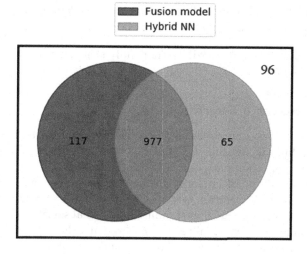

Fig. 4. Venn diagram of the correct predictions

4 Discussion

4.1 Easy Instances and Hard Instances

Figure 4 shows the overlap of the Fusion model and the Hybrid NN mdoel. Both models can correctly predict 977 instances in the test set. As we discussed, the Fusion model is good at general (sub)categories, and the Hybrid model performs better in domain-specific (sub)categories. Those result in the differences between the predictions of both models. Total 96 instances cannot be solved by both models. We find that both models tend to classify the large number into money subcategory in Monetary category. For example, 75,000 in (T2) and 45,176,392 in (T3).

We also find that some errors may be caused by the complex sentence as 2365 in (T4). This kind of instances is not only hard for models, but also difficult for those without domain knowledge. Furthermore, models seem to perform poor for the numerals linked by "-". The 5–7 in (T5) and 24–25 in (T6) are the examples.

(**T2**) \$VFF.CA strong possibility that \$EMH may be the next to uplist toTSX with Village farms deal. 1.1 Million sqft making **75,000** kgs. \$WEED.CA

(**T3**) \$VRX Holy crap short interest increased 9,750,067 shares to **45,176,392** during the last half of October. Some hedgies are in deep doo doo

(**T4**) \$SPY Bought \$ES_F **2365**/2345 Call Spread @ 3.25. Expires this Friday. Trying to sell it at 8.00 day order.

(T5) \$RAD trash..clearance all b.s..after asset sale 250mill profit AFTER caped puts it **5–7**...dumb azzes

(T6) \$MOMO Big day yesterday so hopefully this morning was just a gap filled from **24–25**.

4.2   Future Research Directions

In order to analyze the numerals in the financial social media data in depth, only understanding the meanings of numerals themselves is not enough. Because there may have more than one cashtag in a financial tweet. Understanding their semantic roles in the financial social media data is needed when mining fine-grained opinions toward a certain target. Along this line, we design another novel task for fine-grained numeral understanding in financial social media data, called numeral attachment [4], which aims to detect the attached target (i.e., cashtag) of the numeral. In other words, we attempt to understand which cashtag a given numeral is attached to in a tweet. For example, there are two cashtags and one numeral in (T7). The numeral "36.50" is related to \$BEXP, instead of \$KOG.

(T7) \$KOG Took a small position- hopefully a better outcome than getting kneecapped by \$BEXP selling itself dirt cheap at **36.50**

Previous works mentioned that we should not only analyze the numerals with context, but also need to link the numerals with the outside resources [6, 8]. This is also one of the important challenges when analyzing numerals in textual data. Some tables in formal documents such as the 10-K report can be seen as the outside resource for the financial tweets. Ibrahim and Weikum [9] demonstrated a system to link the numeral information in the documents with table data. This kind of task is more challenging with financial social media data, since the description in financial tweets are informal and short. Furthermore, the numeral information such as market price can be used for generating market comments [11] and capturing the interest of the social trading platform users [3]. It also shows the importance of referring to the market price when analyzing the financial textual data.

5   Conclusion

According to World Economic Forum 2015, social trading is one of the crucial trends in FinTech tide. In the FinNum task series, We presented novel and important issues in analyzing the numerals in financial social media data in a fine-grained way, and provided a large and high quality dataset to lead the new track of numeral understanding. The proposed tasks are the vanguard of in-depth opinion mining for financial social media data.

In this paper, we analyze the state-of-the-art models in depth, and sort out the strength of each model. The results can provide some insights for future works, which also deal with the numeral data. Several challenges of fine-grained understanding toward numerals in financial social media data and financial textual data are also listed.

For the next edition of FinNum task, we will present the annotations related to numeral attachment of financial social media data.

Acknowledgments. We greatly appreciate the efforts of all the participants in the FinNum shared task at NTCIR-14.

This shared task was partially supported by the Ministry of Science and Technology, Taiwan, under grants MOST-106-2923-E-002-012-MY3, 108-2218-E-009-051-, MOST-108-2634-F-002-008-, and MOST 107-2218-E-009-050-, and by Academia Sinica, Taiwan, under grant AS-TP-107-M05.

References

1. Ait Azzi, A., Bouamor, H.: Fortia1@ the NTCIR-14 FinNum task: enriched sequence labeling for numeral classification. In: Proceedings of the 14th NTCIR Conference on Evaluation of Information Access Technologies (2019)

2. Chen, C.C., Huang, H.H., Chen, H.H.: Crowd view: converting investors' opinions into indicators. In: Proceedings of the 28th International Joint Conference on Artificial Intelligence (IJCAI) (2019)

3. Chen, C.C., Huang, H.H., Chen, H.H.: Next cashtag prediction on social trading platforms with auxiliary tasks. In: 2019 IEEE/ACM International Conference on Advances in Social Networks Analysis and Mining (ASONAM) (2019)

4. Chen, C.C., Huang, H.H., Chen, H.H.: Numeral attachment with auxiliary tasks. In: The 42nd International ACM SIGIR Conference on Research & Development in Information Retrieval. ACM (2019)

5. Chen, C.C., Huang, H.H., Shiue, Y.T., Chen, H.H.: Numeral understanding in financial Tweets for fine-grained crowd-based forecasting. In: 2018 IEEE/WIC/ACM International Conference on Web Intelligence (WI), pp. 136–143. IEEE (2018)

6. Chen, C.C., Huang, H.H., Takamura, H., Chen, H.H.: Numeracy-600k: learning numeracy for detecting exaggerated information in market comments. In: Proceedings of the 57th Annual Meeting of the Association for Computational Linguistics (2019)

7. Chen, C.C., Huang, H.H., Tsai, C.W., Chen, H.H.: CrowdPT: summarizing crowd opinions as professional analyst. In: The World Wide Web Conference, WWW '19, pp. 3498–3502. ACM, New York, NY, USA (2019). https://doi.org/10.1145/3308558.3314122

8. Ibrahim, Y., Riedewald, M., Weikum, G., Zeinalipour-Yazti, D.: Bridging quantities in tables and text. In: 2019 IEEE 35th International Conference on Data Engineering (ICDE), pp. 1010–1021. IEEE (2019)

9. Ibrahim, Y., Weikum, G.: Exquisite: explaining quantities in text. In: The World Wide Web Conference, WWW '19, pp. 3541–3544. ACM, New York, NY, USA (2019). https://doi.org/10.1145/3308558.3314134

10. Liang, C.C., Su, K.Y.: ASNLU at the NTCIR-14 FinNum task: incorporating knowledge into DNN for financial numeral classification. In: Proceedings of the 14th NTCIR Conference on Evaluation of Information Access Technologies (2019)

11. Murakami, S., et al.: Learning to generate market comments from stock prices. In: Proceedings of the 55th Annual Meeting of the Association for Computational Linguistics (Volume 1: Long Papers), pp. 1374–1384. Association for Computational Linguistics, Vancouver, Canada, July 2017. https://doi.org/10.18653/v1/P17-1126, https://www.aclweb.org/anthology/P17-1126

12. Spark, A.: BRNIR at the NTCIR-14 FinNum task: scalable feature extraction technique for numeral classification. In: Proceedings of the 14th NTCIR Conference on Evaluation of Information Access Technologies (2019)

13. Tian, K., Peng, Z.J.: AIAI at the NTCIR-14 FinNum task: financial numeral tweets fine-grained classification using deep word and character embedding-based attention model. In: Proceedings of the 14th NTCIR Conference on Evaluation of Information Access Technologies (2019)

14. Wang, W., Liu, M., Zhang, Z.: WUST at the NTCIR-14 FinNum task. In: Proceedings of the 14th NTCIR Conference on Evaluation of Information Access Technologies (2019)

15. Wu, Q., Wang, G., Zhu, Y., Liu, H., Karlsson, B.: DeepMRT at the NTCIR-14 FinNum task: a hybrid neural model for numeral type classification in financial tweets. In: Proceedings of the 14th NTCIR Conference on Evaluation of Information Access Technologies (2019)

Financial Numeral Classification Model Based on BERT

Wei Wang[1,2], Maofu Liu[1,2(✉)], Yukun Zhang[1,2], Junyi Xiang[1,2], and Ruibin Mao[3]

[1] School of Computer Science and Technology,
Wuhan University of Science and Technology, Wuhan 430065, China
liumaofu@wust.edu.cn
[2] Hubei Province Key Laboratory of Intelligent Information Processing and Real-Time Industrial System, Wuhan 430065, China
[3] Center for Studies of Information Resources,
Wuhan University, Wuhan 430072, China

Abstract. Numerals contain rich semantic information in financial documents, and they play significant roles in financial data analysis and financial decision making. This paper proposes a model based on the Bidirectional Encoder Representations from Transformers (BERT) to identify the category and sub-category of a numeral in financial documents. Our model holds the obvious advantages in the fine-grained numeral understanding and achieves good performance in the FinNum task at NTCIR-14. The FinNum task is to classify the numerals in financial tweets into seven categories, and further extend these categories into seventeen subcategories. In our proposed model, we first analyze the obtained financial data from the FinNum task and enhance data for some subcategories by entity replacement. And then, we adopt our fine-tuning BERT model to finish the task. As a supplement, some popular traditional and deep learning models have been selected for comparative experiments, and the experimental results show that our model has achieved the state-of-the-art performances.

Keywords: Financial numeral classification · Financial data processing · FinNum task · BERT model

1 Introduction

FinTech (Financial Technology) [1] has become a hot topic that attracts much attention in recent years. An increasing number of investors prefer to share their investment skills, stock reviews and investment recommendations through various social platforms, and moreover, social media data also contains much rich financial information. Due to the informal writing style, social media data is also unstructured and noise-induced, which holds more challenges than news and official document analysis.

In the financial field, the numeral is a crucial part of financial documents. In order to understand the detail of opinions in the financial documents, we should not only analyze the text but also need to assay the numeric information at semantic level [2],

M. P. Kato et al. (Eds.): NTCIR 2019, LNCS 11966, pp. 193–204, 2019.
https://doi.org/10.1007/978-3-030-36805-0_15

and the fine-grained classification [3] of numerals is necessary. The numeral classification is a novel issue that needs to predict the category and subcategory of a specific numeral in free-style text. The commonly used classification models include Naive Bayes [4], CNN (Convolutional Neural Networks) [5], RNN (Recurrent Neural Network) [6] and LSTM (Long Short-Term Memory) [7]. In recent years, BERT (Bidirectional Encoder Representations from Transformers) [8], a language representation model proposed by Google, has achieved significant improvements in a wide variety of natural language processing tasks after fine-tuning, and hence, we put forward a model based on BERT for the FinNum [9] task at NTCIR-14 to identify the category and subcategory of a numeral in financial documents.

The BERT is a pre-trained language representation model, meaning that it trains a universal language understanding model on a large-scale corpus, e.g. Wikipedia [10]. The pre-trained model provides powerful context-dependent sentence representation for downstream tasks in NLP, e.g. Question Answering (QA), Named Entity Recognition (NER). The BERT outperforms previous methods because it is the first unsupervised, deeply bidirectional system for pre-trained NLP [11].

In FinNum task, numerals are the objects of classification. The FinNum task contains Task1 and Task2, and the Task1 is main category classification, which is to classify the numerals in financial tweets into seven categories, i.e. Indicator, Monetary, Option, Percentage, Product Number, Quantity and Temporal. On the basis of the Task1, the Task2 further extends the seven categories to seventeen subcategories. We adjust the input module based on BERT, and the model can automatically recognize the category labels.

The rest of this paper is organized as follows. Section 2 shows the related work of numeral classification. Section 3 describes our classification model based on BERT. Section 4 discusses the official experimental results from NTCIR-14 and our additional experimental results. Finally, we draw some conclusions in Sect. 5.

2 Related Work

The text classification [12, 13] is a common problem in NLP. In view of large-scale datasets, Zhang et al. [14] proposed a character-level convolutional neural network for text classification, showing that CNN can be directly applied to the distributed or discrete embedding of words, without any knowledge on the syntactic or semantic structures of a language. Lai et al. [15] attempted to use RNN for text classification when getting rid of human-designed features, and they applied a recurrent structure to capture contextual information to learn word representations as far as possible. The experiments results showed that RNN model also had a good performance in text classification, particularly on document-level data sets. Some researchers also proposed other methods of classification, such as attention based CNN [16], hierarchical attention networks [17], attention based LSTM [18], adversarial multi-task learning [19].

To address the challenge of data sparsity, a variety of methods have been put forward for training general purpose language representation models using an enormous amount of unannotated texts, such as ELMo [20], Generative Pre-trained Transformer (GPT) [21] and BERT. The pre-trained models can be fine-tuned on NLP

tasks and have achieved significant improvement over training on task-specific annotated data. In 2019, Chen et al. [22] adopted BERT for joint intent classification and
slot filling and proved that the BERT model is available for the classification task.

There are a few specific studies on the classification of financial texts. In 2009,
Schumaker and Chen [23] proposed the AZFinText system to predict the breaking
financial news from 9,211 financial news articles. In their experiments, the Support
Vector Machine (SVM) [24] model was used to perform a binary classification in two
predefined categories, i.e. stock price rise and drop.

In financial statements, numerals are usually presented in a structured form that is
easier to classify. However, numerals in social media data are always unstructured and
noise-induced. In order to capture the point of an investor, to understand the numerals
at a semantic level is indispensable for fine-grained numeral classification. In this
paper, we propose a financial numeral classification model based on BERT.

3   Classification Model Based on BERT

Our model includes three main modules, i.e. data preprocessing, feature extraction and
classifier, and Fig. 1 illustrates the overview of our proposed model in detail.

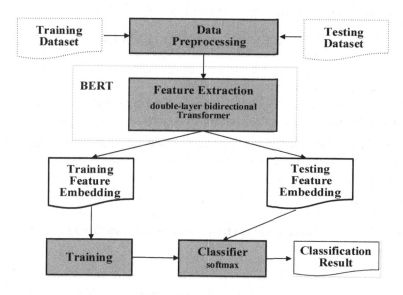

Fig. 1. Model overview.

3.1   Data Preprocessing

3.1.1   Overall Extension

If a tweet contains multiple target number, we expand this tweet into multiple data,
ensuring that one target number corresponds to one tweet and one category. As shown
in Example 1 below.

Example 1:
Original tweet:
Target number: 76.33, 6
Tweet: I am bearish on $RACE with a target price of $76.33 in 6 mos. on Vetr!
Expanded tweet:
Target number: 76.33
Tweet: I am bearish on $RACE with a target price of $76.33 in 6 mos. on Vetr!
Target number: 6
Tweet: I am bearish on $RACE with a target price of $76.33 in 6 mos. on Vetr!
In the first-round experiment, we choose the method mentioned above for data expansion. Considering that other numbers in the tweet have noise influence on the Target number, we carry out fine-grained processing on the tweet in the second-round experiment, and only the Target number is retained, as shown in Example 2.

Example 2:
Original tweet:
Target number: 76.33, 6
Tweet: I am bearish on $RACE with a target price of $76.33 in 6 mos. on Vetr!
Expanded tweet:
Target number: 76.33
Tweet: I am bearish on $RACE with a target price of $76.33 in mos. on Vetr!
Target number: 6
Tweet: I am bearish on $RACE with a target price of $ in 6 mos. on Vetr!

3.1.2   Partial Extension
In Task1, due to the uneven data distribution, we need to expand the data of Indicator, Option and Product, We have tried two different ways to partially enhance the data.

Splitting: Dividing tweets with the target number and adding the segmented data into the training set, as a result, the data of these three categories is tripled.

Example 3:
Original tweet:
Target number: 3
Tweet: $ACAD Nelotanserin will not be on the market for another 3 years. And that is assuming best-case scenario.
Splitting:
Target number: 3
Tweet: $ACAD Nelotanserin will not be on the market for another **3**
Tweet: **3** years. And that is assuming best-case scenario.

Entity Replacement. We find that a large number of tweets contain stock codes in the form of "$XXX", so we extract all these codes as entities, and then randomly select a code from these entities and replace the stock codes in the original tweet, and thus we get a new data. As shown in Example 4.

Example 4:

Original tweet:

Target number: 3

Tweet: **$ACAD** Nelotanserin will not be on the market for another 3 years. And that is assuming best-case scenario.

Entity Replacement:

Target number: 3

Tweet: **$TSLA** Nelotanserin will not be on the market for another 3 years. And that is assuming best-case scenario.

However, in Task1, the ablation experimental results from Table 1 show that the two data expansion methods mentioned above reduce the accuracy of the validation set. Therefore, we do not expand the experimental data in Task1.

Table 1. The ablation experiments of Partial Extension.

	Task1	
	Micro-F1	Macro-F1
Splitting	0.9302	0.8611
Entity replacement	0.9416	0.8723
No extension	**0.9505**	**0.9101**

In Task2, we also do the ablation experiment of subcategory-data extension, the first-round experimental results show that expand Monetary, Temporal and Option categories can achieve the best results in the way of Example 3. In the second-round experiments, we find that only expanding the data for Option and Monetary, as the way in Example 4, can obtain the better performance than the other categories.

3.2 Feature Extraction

In our model, the representation of input tweet is a concatenation of token embedding, sentence embedding, and positional embedding, and the representation of the input tweet has been shown in Fig. 2.

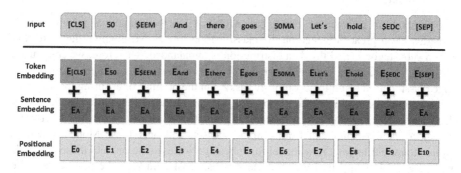

Fig. 2. The representation of the input tweet. The input tweet is "50 $EEM and there goes 50MA, let's hold $EDC".

The feature extraction is performed by two-layer bi-directional Transformer in our model, and the Transformer [25] is an encoder-decoder structure, which has a good performance in feature extraction at the sentence level by relying on the attention mechanism. After the fine-tuning in experimental datasets, the label embedding is contained in the [CLS] token of the Transformer layers' output, as shown in Fig. 3.

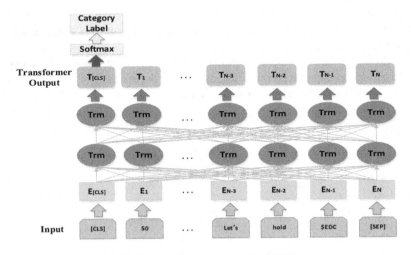

Fig. 3. Feature extraction using BERT.

The BERT uses a simple approach in the pre-trained stage, in which a special classification embedding ([CLS]) inserted as the first token and a special token ([SEP]) added as the final token. It masks out 15% of the words in the input, runs the entire sequence through a deep bidirectional Transformer encoder, As shown in Example 5.

Example 5:
Tweet:[CLS] 50 $EEM And there [MASK1] 50MA, let's [MASK2] $EDC [SEP]
    Masked words: [MASK1] = goes; [MASK2] = hold.

3.3   Classifier

The numerals in a tweet are classification objects, so the contextual information of numerals has a great influence on the classification model. Meanwhile, there are seventeen subcategories and the data distribution is unbalanced. Therefore, we construct a two-stage classification model, which firstly deals with Task1 and then use the first-stage classification result to accomplish Task2. On the premise of ensuring the accuracy of the main classification as far as possible, we carry out the fine-grained subcategory classification. For example, if a numeral is predicted as Monetary by the Task1, then we select all Monetary data from training dataset, and only these Monetary data is applied to retrain to get a new model for subcategory classification, and those numerals predicted as Monetary in Task1 will accomplish Task2 through the new model. We ensure that the main category and subcategory are consistent in this two-stage model. The details are shown in Fig. 4.

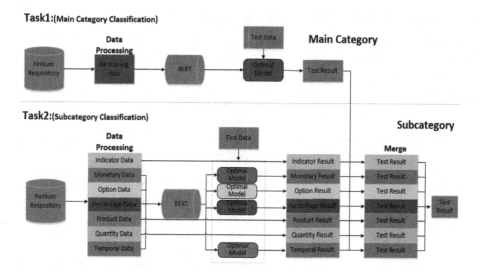

Fig. 4. The two-stage classification model.

In the two-stage classification model, we employ the softmax function to accomplish the final classification, and the details are shown in Fig. 3. Based on the first special token ([CLS]) of the Transformer output, denoted h_1, the class label is predicted by the following formula (1),

$$y^i = \text{softmax}(W^i h_1 + b^i) \tag{1}$$

where y^i is the class label.

Considering that our classification model needs only one input rather than a pair of sentences, we simply adjust the input, after that, the code can automatically define the label list according to the number of categories, without manually setting, which can adapt to a variety of multi-classification tasks.

4 Experiments

4.1 Settings

Our experimental data is from the FinNum task, which has been crawled from Stocktwits. Before the experiments, we conduct statistics on 6860 training data after Overall Extension, Table 2 illustrates the distribution of main categories and subcategories, and the data is not an even distribution. Because the Monetary and Temporal categories take a large proportion, accounting for 35.96% and 34.56% respectively, and while the data of Indicator, Option, and Product Number takes a small proportion, accounting for 2.43%, 2.46%, and 1.66% respectively.

Table 2. The distribution of categories and subcategories

Category	Number of instances	Percentage (%)	Subcategory	Number of instances	Percentage (%)
Indicator	167	2.43	Indicator	167	2.43
Monetary	2467	35.96	Buy price	319	4.65
			Change	143	2.08
			Forecast	270	3.94
			Money	589	8.59
			Quote	792	11.55
			Sell price	103	1.50
			Stop loss	25	0.36
			Support or resistance	226	3.29
Option	169	2.46	Exercise price	113	1.65
			Maturity date	56	0.82
Percentage	838	12.22	Absolute	253	3.69
			Relative	585	8.53
Product Number	114	1.66	Product number	114	1.66
Quantity	741	10.80	Quantity	741	10.80
Temporal	2364	34.56	Date	2079	30.31
			Time	285	4.15

In this paper, we adopt the common evaluation metrics in a multi-classification task, i.e. Micro-F1 and Macro-F1 [26], Precision, Recall, and F1-measure, as shown in the following formulas and descriptions.

$$Precision = \frac{TP}{TP + FP} \tag{2}$$

$$Recall = \frac{TP}{TP + FN} \tag{3}$$

$$F1 - measure = \frac{2 * Precision * Recall}{Precision + Recall} \tag{4}$$

Macro-F1: Calculated globally by counting the total true positives, false negatives and false positives.

Micro-F1: Calculated by counting the true positives, false negatives and false positives of each class.

Among them, the TP (True Positives), FP (False Positives), and FN (False Negatives) refer to the category where cases are truly classified into the positive class, wrongly classified into the positive class, and wrongly classified into the positive class separately.

4.2 Results

As a supplement, some popular traditional and deep learning models, e.g. SVM, TextCNN, BiLSTM, are adopted for comparative experiments, and the experimental results are listed in Table 3.

Table 3. The evaluation results of different models.

Model	Task1		Task2	
	Micro-F1	Macro-F1	Micro-F1	Macro-F1
SVM	0.7402	0.6371	0.6088	0.5293
TextCNN	0.7323	0.5996	0.6207	0.5507
BiLSTM	0.7633	0.5933	0.5418	0.3643
BiLSTM+Attention	0.7928	0.7023	0.6876	0.6190
BERT (first-round)	0.9450	0.8862	0.8725	0.8307
BERT (second-round)	**0.9505**	**0.9101**	**0.8972**	**0.8798**

From the experimental results, we can see that both the SVM model and deep learning models have achieved good performance. However, there is still much room for improvement in the experimental results. Both Task1 and Task2 have greatly improved in the results with our classification model based on BERT.

In FinNum Task, we submitted one classification result by SVM, in additional experiments, our proposed model achieves better results. The official evaluation results are listed in Table 4.

Table 4. The official evaluation results.

Team	Task1		Task2	
	Micro-F1	Macro-F1	Micro-F1	Macro-F1
Stark	0.7801	0.6175	0.6908	0.5683
BRNIR	0.7427	0.6353	0.6199	0.4714
WUST (SVM)	**0.7402**	**0.6371**	**0.6088**	**0.5293**
aiai	0.8645	0.7809	0.8024	0.7411
Fortia2	0.8988	0.7926	0.7705	0.6886
ZHAW	0.8645	0.7927	0.7554	0.6644
ASNLU	0.8972	0.8093	0.7912	0.7251
DeepMRT	0.9187	0.8794	0.8303	0.7790
Fortia1	0.9394	0.9005	0.8717	0.8240

WUST is our team name, and from Tables 3 and 4. We can find that, in the second-round experiments, our proposed model has achieved the best performances on both Task1 and Task2.

4.3 Discussions

According to the classification results of BERT and SVM models, we have analyzed several examples of classification errors and discussed the reasons for these errors. The classification result of BERT is better than that of SVM, and therefore, we first analyze some cases where BERT classification is correct but SVM classification is wrong, which are listed as follows.

(1) In Example 6, for the target number "85", the SVM classification result is Temporal, while the BERT classification result is correct and the category should be Monetary.

Example 6:
Target number: 85
    Tweet: 85 $QD post net income of 85 M on total rev. of M In six months that ended June of this year, rev. was M &net income came at M.
    The target number is "85", and the original tweet is "2016 $QD post net income of 85M on total rev. of 212M In six months that ended June 30 of this year, rev. was 270M & net income came at 143M". In this tweet, the word corresponding to "85" is "85M". The SVM model can not recognize "85" as the keyword, resulting in SVM classification error. For BERT model, positional information is added in feature extraction, which can easily identify the correct category according to the contextual information.

(2) In Example 7, for target number 4, the SVM classification result is Percentage, while the BERT classification result is correct and the category should be Product.

Example 7:
Target number: 4
    Tweet: $AMD get in before mainstream gets the news. Nov NPD just released! Xbox and ps4 up % YOY.pro and X sell for big profits.
    The target number is "4", and the original tweet is "$AMD get in before mainstream gets the news. Nov NPD just released! Xbox and ps4 up 40% YOY.pro and X sell for big profits". In this tweet, both the words "ps4" and "40%" contain the number "4", while "%" is a strong feature in Percentage category, which misleads the SVM to classify target number as Percentage for these reasons, and however, the BERT model can accurately identify the keyword corresponding to the target number as "ps4", and correctly classify the target number "4" as Product category according to the "Xbox" in the contextual information.
    From the Examples 6 and 7, compared with SVM model, we can draw a conclusion that the BERT model has more advantages in identifying a keyword in the tweet and makes good improvement in the classification task because it retains positional information when extracting textual features.

Although the classification results are excellent by BERT model, there are still some drawbacks, which are listed as follows.

(3) In Example 8, for target number "60", the BERT classification result is Percentage, while the correct category should be Product.

Example 8:
Target number: 60

Tweet: $LODE just gonna keep watching all of these, and try to add more to lode, goal 60k shs.

The target number is "60", and the original tweet is "$LODE just gonna keep watching all of these, and try to add more to lode, goal 60k shs.". When the target number is in the end part of tweet, our model cannot extract the contextual information correctly. As shown in Example 8, the content after the target number is just only a word "shs", leading to a misclassification into the Monetary category, but the correct category should be Quantity.

5   Conclusions

In this paper, in order to understand the rich semantic information contained in numerals in financial documents, we propose a financial numeral classification model based on BERT, which makes numerals in financial texts can be easily and correctly classified. Our model has achieved significant performance in FinNum task at NTCIR-14.

In our classification model, we only use BERT model for feature extraction. In the future, we will consider employing a joint model for numeral classification, e.g. BERT and BiLSTM. We plan to improve our model through error analysis and language phenomena in financial text. Moreover, we would like to extract and select more rules and semantic features for our model to improve classification accuracy.

Acknowledgments. The work presented in this paper is partially supported by the Major Projects of National Social Foundation of China under Grant No. 11&ZD189.

References

1. Dhar, V., Stein, R.M.: FinTech platforms and strategy. Commun. ACM **60**(10), 32–35 (2017)
2. Zhou, Y., Ni, B., Yan, S., Moulin, P., Tian, Q.: Pipelining localized semantic features for fine-grained action recognition. In: Fleet, D., Pajdla, T., Schiele, B., Tuytelaars, T. (eds.) ECCV 2014. LNCS, vol. 8692, pp. 481–496. Springer, Cham (2014). https://doi.org/10.1007/978-3-319-10593-2_32
3. Karaoglu, S., et al.: Con-text: text detection for fine-grained object classification. IEEE Trans. Image Process. **26**, 3965–3980 (2017)
4. McCallum, A., Nigam, K.: A comparison of event models for Naive Bayes text classification. In: AI-98 Workshop on Learning for Text Categorization, vol. 752, no. 1, pp. 41–48 (1998)
5. Kim, Y.: Convolutional neural networks for sentence classification. Eprint Arxiv (2014)

6. Hüsken, M., Stagge, P.: Recurrent neural networks for time series classification. Neurocomputing **50**, 223–235 (2003)
7. Zhou, C., Sun, C., et al.: A C-LSTM neural network for text classification. Comput. Sci. **1**(4), 39–44 (2015)
8. Devlin, J., Chang, M.W., Lee, K., et al.: BERT: pre-training of deep bidirectional transformers for language understanding. arXiv preprint arXiv:1810.04805 (2018)
9. Chen, C.C., Huang, H.H., Takamura, et al.: Overview of the NTCIR-14 FinNum task: fine-grained numeral understanding in financial social media data. In: Proceedings of the 14th NTCIR Conference on Evaluation of Information Access Technologies (2019)
10. Zhu, Y., Ryan, K., Richard, S., et al.: Aligning books and movies: towards story-like visual explanations by watching movies and reading books. In: 2015 IEEE International Conference on Computer Vision, pp. 19–27 (2015)
11. Lee, et al.: BioBERT: pre-trained biomedical language representation model for biomedical text mining. arXiv preprint arXiv:1901.08746 (2019)
12. Armand, J., Edouard, G., Piotr, B., Douze, M., et al.: FastText.zip: compressing text classification models. arXiv preprint arXiv:1612.03651 (2016)
13. Moraes, R., Valiati, J.F., Neto, W.P.G.: Document-level sentiment classification: an empirical comparison between SVM and ANN. Expert Syst. Appl. **40**(2), 621–633 (2013)
14. Zhang, X., Zhao, J.B., et al.: Character-level convolutional networks for text classification. In: International Conference on Neural Information Processing Systems (2015)
15. Lai, S., Xu, L., Liu, K., et al.: Recurrent convolutional neural networks for text classification. In: Twenty-Ninth AAAI Conference on Artificial Intelligence (2015)
16. Liu, Y., et al.: An attention-gated convolutional neural network for sentence classification. CoRR (2018)
17. Pappas, N., Popescu, B.A.: Multilingual hierarchical attention networks for document classification. arXiv preprint arXiv:1707.00896 (2017)
18. Zhang, Y., et al.: A text sentiment classification modeling method based on coordinated CNN-LSTM-attention model. Chin. J. Electron. **28**(01), 124–130 (2019)
19. Liu, P.F., Qiu, X., Huang, X.: Adversarial multi-task learning for text classification. arXiv preprint arXiv:1704.05742 (2017)
20. Peters, M.E., et al.: Deep contextualized word representations. arXiv preprint arXiv:1802.05365 (2018)
21. Alec, R., Karthik, N., Tim, S., et al.: Improving language understanding with unsupervised learning. Technical report. OpenAI (2018)
22. Chen, Q., Zhuo, Z., Wang, W.: BERT for joint intent classification and slot filling. arXiv preprint arXiv:1902.10909 (2019)
23. Schumaker, R.P., Chen, H.: Textual analysis of stock market prediction using breaking financial news. The AZFin text system (2009)
24. Suykens, J.A., Vandewalle, J.: Least squares support vector machine classifiers. Neural Process. Lett. **9**(3), 293–300 (1999)
25. Vaswani, A., et al.: Attention is all you need. Advances in Neural Information Processing Systems, pp. 5998–6008 (2017)
26. Yang, Y.: An evaluation of statistical approaches to text categorization. Inf. Retrieval **1**(1–2), 69–90 (1999)

Author Index

Printed in the United States
By Bookmasters